CALCULUS

FOR ENGINEERING I

Second Edition

Colton Solomon
T-Hall 460

CHESTER MIRACLE

Kendall Hunt
publishing company

Cover image © Shutterstock, Inc.

www.kendallhunt.com
Send all inquiries to:
4050 Westmark Drive
Dubuque, IA 52004-1840

Contents

6310 Limits for Derivatives

Suppose we must find the limit of the function $f(x)$ as x approaches b. We write this as

$$\lim_{x \to b} f(x).$$

What does this statement mean? It means to determine what fixed number the numbers $f(x)$ get closer and closer to as we substitute numbers for x that are getting closer and closer to the number b. Suppose the numbers $f(x)$ get closer to the fixed number L as we substitute for x numbers that are getting closer and closer to b. The number L is called the limit. One important part of the process is: we do not actually substitute $x = b$ for x. We substitute the values of x that get closer and closer to b.

Example 1. Find $\lim\limits_{x \to 3} \dfrac{x^2 + 4x}{x + 1}$.

We look to see what happens as we let x take on values such as 3.1, 3.01, 3,001, 3.0001, etc. or 2.9, 2.99, 2.999, etc. Note that these value get closer and closer to 3. We do not use $x = 3$.

When $x = 3.1$, the value of the fraction is 5.3683

When $x = 3.01$, the value of the fraction is 5.2619.

When $x = 3.001$, the value of the fraction is 5.2512.

When $x = 3.0001$, the value of the fraction is 5.2501.

What happens when we use values of x that are less than 3?

When $x = 2.9$ the value of the fraction is 5.1308.

When $x = 2.99$ the value of the fraction is 5.2381.

When $x = 2.999$ the value of the fraction is 5.2488.

When $x = 2.9999$ the value of the fraction is 5.2499.

Continuing in this way we see that the values of the fraction get closer and closer to 5.25 as the values of x get closer and closer to 3. Therefore,

$$\lim_{x \to 3} \frac{x^2 + 4x}{x + 1} = 5.25.$$

After doing several problems like this we discover that there is a great short cut to finding limits. In order to find the limit we substitute values of x that get closer and closer to 3 but do not substitute 3 for x. But suppose we do what is not part of the process of finding the limit. Suppose we replace x by 3. We discover that if we replace x by 3 this gives us the value of the limit.

$$\frac{(3)^2 + 4(3)}{3 + 1} = 5.25.$$

We discover that we can find the limit by just substituting. No need to substitute a lot of numbers that get closer and closer to 3. This substitution method for finding limits is easy. Substitution gives the correct value of the limit because the function $\dfrac{x^2 + 4x}{x + 1}$ is continuous at $x = 3$. We usually define a function $f(x)$ to be continuous at $x = b$ if and only if $\lim\limits_{x \to b} f(x) = f(b)$. This means that the following theorem is true.

Theorem 1. *If $f(x)$ is continuous at $x = b$, then*

$$\lim_{x \to b} f(x) = f(b).$$

All of our elementary functions are continuous for all values of x for which they are defined. This says that the functions e^x, $\sin(x)$ and $\cos(x)$ are continuous for all values of x. Polynomials $P_n(x)$ are continuous for all values of x. The function $\ln(x)$ is continuous for $x > 0$. The quotient of two polynomials $P(x)/Q(x)$ is continuous except for values of x such that the denominator $Q(x) = 0$. The function $\tan x$ is continuous except when x is a multiple of π. Theorem 1 makes it easy to find $\lim\limits_{x \to b} f(x)$ whenever $f(x)$ is continuous. We can find limits involving almost every function we know by just using substitution.

Example 2. Find $\lim\limits_{x \to 3} \dfrac{x^3 - 5x + 4}{x^2 + 10}$.

Solution. Since this function is continuous at $x = 3$, we use Theorem 1 to find the limit. We replace x by 3 and find $\frac{3^3 - 5(3) + 4}{3^2 + 10} = \frac{16}{19}$. Theorem 1 tells us that

$$\lim_{x \to 3} \frac{x^3 - 5x + 4}{x^2 + 10} = \frac{16}{19}.$$

Example 3. Find $\lim\limits_{x\to 5}(e^{3x}+\sin 2x)$.

Solution. The function $e^{3x}+\sin 2x$ is continuous at $x=5$. Substituting 5 for x we get $e^{15}+\sin 10$. Theorem 1 tells us that

$$\lim_{x\to 5}(e^{3x}+\sin 2x)=e^{15}+\sin 10.$$

At this stage in our study of functions we encounter just two kinds of functions that are not continuous. One kind is a function with a jump discontinuity. We will discuss these later. A function has the other kind of discontinuity at $x=b$ whenever $f(x)$ is not defined at $x=b$. This usually means that the denominator of the function is zero at $x=b$. Let us now look at some examples where the function $f(x)$ has a zero in the denominator at $x=b$ and where we are trying to find $\lim\limits_{x\to b} f(x)$. In this situation $f(x)$ is not continuous at $x=b$. Recall that $f(b)$ does not need to be defined in order for $\lim\limits_{x\to b} f(x)$ to exist. Let us now look at a situation where the function is not continuous at the limiting point. Theorem 1 does not work for such cases.

Example 4. Find $\lim\limits_{x\to 2}\dfrac{3x^2-5x-2}{x^2-4}$.

Solution. First, we try the idea which works on all continuous functions. We substitute 2 for x to get

$$\lim_{x\to 2}\frac{3x^2-5x-2}{x^2-4}=\frac{3\cdot 4-5\cdot 2-2}{4-4}=\frac{0}{0}.$$

The substitution does not give a number. Theorem 1 can not be applied because the function is not continuous at $x=2$. Division by zero is the only arithmetical operation that is not defined in our number system. The expression $\frac{0}{0}$ is usually called an indeterminate form. It can not be determined. We have also discovered that the function as given is not continuous at $x=2$ because it is not defined for $x=2$. We can not evaluate the limit using Theorem 1, but this does not mean that the limit does not exist.

Let us look at the function in this problem in a little more detail.

$$\frac{3x^2-5x-2}{x^2-4}=\frac{(x-2)(3x+1)}{(x-2)(x+2)}.$$

3

It is true that

$$\frac{3x^2 - 5x - 2}{x^2 - 4} = \frac{3x + 1}{x + 2} \quad \text{when } x \neq 2 \text{ and } x \neq -2.$$

The fraction on the left side of the equal is not defined for $x = 2$ and $x = -2$. The fractions are equal except when $x = 2$ and $x = -2$. The fractions can not be equal for $x = 2$ and $x = -2$ because the fraction on the left is not defined for these values of x. First, using Theorem 1 on the simple fraction on the right side of the equal sign, we get

$$\lim_{x \to 2} \frac{3x + 1}{x + 2} = \frac{7}{4}.$$

We can use Theorem 1 to find this limit since this fraction is continuous at $x = 2$. Recall that we never substitute the value $x = 2$ when finding the limit as x approaches 2. If we substitute any number close to 2 but not 2, then

$$\frac{3x^2 - 5x - 2}{x^2 - 4} \quad \text{and} \quad \frac{3x + 1}{x + 2}$$

have the same value. These two fractions have the same value when $x = 2.1$, or $x = 2.01$, or $x = 2.001$. They also have the same value when $x = 1.9, 1.99$. These are the numbers we would substitute for x when finding the limit as $x \to 2$. Since these two fractions have the same value when we substitute numbers for x that are near 2 it follows that these two fractions have the same limit as x approach 2.

$$\lim_{x \to 2} \frac{3x^2 - 5x - 2}{x^2 - 4} = \lim_{x \to 2} \frac{3x + 1}{x + 2}.$$

We usually write the solution of this limit problem in the shorter form as follows:

$$\lim_{x \to 2} \frac{3x^2 - 5x - 2}{x^2 - 4} = \lim_{x \to 2} \frac{(x - 2)(3x + 1)}{(x - 2)(x + 2)} = \lim_{x \to 2} \frac{3x + 1}{x + 2} = \frac{7}{4}.$$

Example 5. Find $\displaystyle\lim_{x \to -3} \frac{2x^3 + 6x^2 - 5x - 15}{x^2 + x - 6}$.

Solution. With high hopes we try to use Theorem 1 to find the limit by substituting -3 for x.

$$\frac{2(-3)^3 + 6(-3)^2 - 5(-3) - 15}{(-3)^2 + (-3) - 6} = \frac{0}{0}.$$

This did not give the answer. We cannot divide by zero. We got an indeterminate form. Theorem 1 cannot be applied. Taking another look at the fraction we see that the fraction is not continuous at $x = -3$. We need to factor out the factor $(x + 3)$. We know that $x - (-3)$ is a factor of the denominator because the value of the denominator is zero when we replace x with -3. We know that $x + 3$ must be a factor of the numerator since the value of the numerator is 0 when we replace x by -3. In order to find the other factor we can use long division as follows:

$$
\begin{array}{r}
2x^2 - 5 \\
x + 3 \overline{\smash{\big)}\, 2x^3 + 6x^2 - 5x - 15} \\
\underline{2x^3 + 6x^2 } \\
-5x - 15 \\
\underline{-5x - 15}
\end{array}
$$

This long division tells us that $2x^3 + 6x^2 - 5x - 15 = (x + 3)(2x^2 - 5)$. We can factor $x^2 + x - 6$ as $(x + 3)(x - 2)$.

$$\frac{2x^3 + 6x^2 - 5x - 15}{x^2 + x - 6} = \frac{(x + 3)(2x^2 - 5)}{(x + 3)(x - 2)} = \frac{2x^2 - 5}{x - 2}$$

when $x \neq -3$ and $x \neq 2$. The fractions

$$\frac{2x^3 + 6x^2 - 5x - 15}{x^2 + x - 6} \quad \text{and} \quad \frac{2x^2 - 5}{x - 2}$$

have the same value except for $x = -3$ and $x = 2$. If we substitute a number near -3 for x into the fraction

$$\frac{2x^2 - 5}{x - 2}$$

we would get the same answer that we would get if we substituted the same numbers into the fraction

$$\frac{2x^3 + 6x^2 - 5x - 15}{x^2 + x - 6}.$$

For example, both fractions have the value, namely -2.41, when x is replaced by -2.9. Recall that when finding the limit as x approaches -3 we do not actually substitute in the value -3. This means when finding the limit

$$\lim_{x \to -3} \frac{2x^3 + 6x^2 - 5x - 15}{x^2 + x - 6}$$

the values we get for this fraction would be the same values we use in finding the limit

$$\lim_{x \to -3} \frac{2x^2 - 5}{x - 2}.$$

Therefore,

$$\lim_{x \to -3} \frac{2x^3 + 6x^2 - 5x - 15}{x^2 + x - 6} = \lim_{x \to -3} \frac{2x^2 - 5}{x - 2}.$$

When computing either limit the two fractions have the same value. This is the reason that the limits are equal. We easily find this last limit using Theorem 1 as follows:

$$\lim_{x \to -3} \frac{2x^2 - 5}{x - 2} = \frac{2 \cdot 9 - 5}{-3 - 2} = -\frac{13}{5}.$$

This tells us that

$$\lim_{x \to -3} \frac{2x^3 + 6x^2 - 5x - 15}{x^2 + x - 6} = -\frac{13}{5}.$$

Note that the fraction $\dfrac{2x^2 - 5}{x - 2}$ is not defined when $x = 2$. The number 2 is not close to -3 and so plays no role in finding the limit as x approaches -3.

Example 6. Find $\lim_{x \to 3} \dfrac{x^2 + 1}{x^2 - x - 6}$.

Solution. First, maybe we can apply the theorem. Substituting 3 for x gives

$$\frac{3^2 + 1}{3^2 - 3 - 6} = \frac{10}{0}.$$

Since division by zero is not defined, Theorem 1 cannot be applied. Note that when we substituted we got $10/0$ and not the indeterminate form $0/0$. This means that $(x - 3)$ is not a factor of the numerator. Next, let us try a few values of x close to 3.

When $x = 3.1$, the value of the fraction is 20.8

When $x = 3.01$, the value of the fraction is 200.8

When $x = 3.001$, the value of the fraction is 2000.8

When $x = 2.9$, the value of the fraction is -19.2

When $x = 2.99$, the value of the fraction is -199.2

When $x = 2.999$, the value of the fraction is -1999.2

It is clear that as the value of x gets closer to 3 the value of the fraction gets larger and larger. In this case the limit does not exist. The value of the fraction does not get closer and closer to some particular number as x gets closer to 3. The limit does not exist. In order to indicate that the value of the fraction is getting larger and larger, we often write $\lim\limits_{x \to 3} \dfrac{x^2 + 1}{x^2 - x - 6} = \infty$. This does not mean that ∞ is a number. We are just indicating how the limit fails to exist. Note that writing ∞ without either a plus or minus sign indicates that the absolute value of the fraction gets larger and larger.

Exercises

1. What is the value of each of the following fractions when $x = 2$? When $x = 1$?

$$\frac{x^2 - 7x + 10}{x^2 + 2x - 8} \quad \text{and} \quad \frac{x - 5}{x + 4}.$$

For what values of x are these two fractions equal? For what values of x are these two fractions not continuous?

2. What is the value of each of the following fractions when $x = -4$? When $x = -(1/3)$? When $x = 0$?

$$\frac{2x^2 + 5x - 12}{3x^2 + 13x + 4} \quad \text{and} \quad \frac{2x - 3}{3x + 1}.$$

3. a) What is the value of each of the following fractions when $x = 5$?

$$\frac{2x^2 - 7x - 15}{x^2 - x - 20} \quad \text{and} \quad \frac{2x + 3}{x + 4}.$$

(b) Explain why $\lim_{x \to 5} \dfrac{2x^2 - 7x - 15}{x^2 - x - 20} = \lim_{x \to 5} \dfrac{2x + 3}{x + 4}$.

(c) Find $\lim_{x \to 5} \dfrac{2x^2 - 7x - 15}{x^2 - x - 20}$.

(d) Find $\lim_{x \to -4} \dfrac{2x^2 - 7x - 15}{x^2 - x - 20}$.

(e) Find $\lim_{x \to 3} \dfrac{2x^2 - 7x - 15}{x^2 - x - 20}$.

4. a) Explain why $\lim_{x \to 3} \dfrac{2x^3 - 6x^2 + 3x - 9}{2x^2 - 3x - 9} = \lim_{x \to 3} \dfrac{2x^2 + 3}{2x + 3}$.

b) Find $\lim_{x \to 1} \dfrac{2x^3 - 6x^2 + 3x - 9}{2x^2 - 3x - 9}$.

5. Find $\lim_{x \to 2} \dfrac{x^3 - 5x^2 + 11x - 10}{x^2 + x - 6}$.

6. Find $\lim_{x \to 2} \dfrac{\dfrac{x + 1}{x + 3} - \dfrac{3}{5}}{x - 2}$. Also find $\lim_{x \to -3} \dfrac{\dfrac{x + 1}{x + 3} - \dfrac{3}{5}}{x - 2}$.

7. Find $\lim_{x \to 3} \dfrac{\dfrac{3x + 2}{2x + 5} - 1}{x - 3}$. Also find $\lim_{x \to 1} \dfrac{\dfrac{3x + 2}{2x + 5} - 1}{x - 3}$.

8

6311 Review of Lines

By far the easiest way to describe a line in a plane is to use analytic geometry. This means we think of the plane as a plane with cartesian coordinates and give the equation of the line. If we graph the equation we obtain the line. From algebra we can recall the general equation of a line. The equation of the line through the point with coordinates (x_1, y_1) and slope m is

$$y - y_1 = m(x - x_1).$$

Example 1. Find the equation of the line through the point $(4, 1)$ with slope $1/2$.

Solution. Since the point is (4,1) this means that $x_1 = 4$ and $y_1 = 1$. When the slope is $1/2$ this means $m = 1/2$. Substituting these numbers into the general form we get .

$$y - 1 = 1/2(x - 4).$$

This simplifies to

$$y = (1/2)x - (1).$$

We obtain the graph of $y = (1/2)x - (1)$ as follows. We first find a solution pair for the equation $y = (1/2)x - (1)$. When $x = -4$, we get $y = (1/2)(-4) - (1) = -3$. A solution pair is $(-4, -3)$. We plot this point and it is a point on the graph. When we plot all solution pairs we obtain a line. This line is the graph of $y = (1/2)x - (1)$.

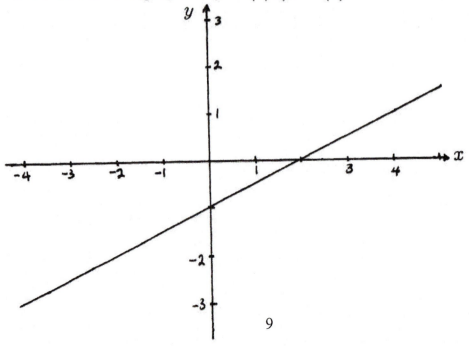

Example 2. Find the equation of the line through the point $(-5, 8)$ with slope $3/2$.

Solution. The line passes through the point $(-5, 8)$ and so $x_1 = -5$ and $y_1 = 8$. For a slope of $3/2$, we use $m = 3/2$. Substituting these numbers into the general form, we get

$$y - 8 = (3/2)(x - (-5)).$$

This simplifies to

$$2y = 3x + 31.$$

When studying geometry we often hear the statement "a line is determined by two points". Suppose we are given the coordinates of two points and wish to determine the equation of the line through these two points. The form given at the beginning of our discussion for the equation of a line uses one point and the slope. When given two points we first need to find the slope m of the line in order to use this form. In general, if a line passes through the two fixed points (x_1, y_1) and (x_2, y_2), then the slope of the line is given by the fraction

$$\text{slope} = m = \frac{y_2 - y_1}{x_2 - x_1}.$$

Example 3. Find the equation of the line through the two points $(-1, 5)$ and $(3, 17)$.

Solution. The two points are $(-1, 5)$ and $(3, 17)$. This means that

$$x_1 = -1, \ y_1 = 5, \ x_2 = 3, \ \text{and} \ y_2 = 17.$$

Substituting these numbers into the formula for the slope of the line, we get

$$\text{slope} = m = \frac{17 - (5)}{3 - (-1)} = \frac{12}{4} = 3.$$

Substituting $x_1 = -1$, $y_1 = 5$, and $m = 3$ into the general formula for the line, we get

$$y - 5 = 3(x - (-1)).$$

The equation of the line through $(-1, 5)$ and $(3, 17)$ is

$$y = 3x + 8.$$

Note that when $x = 3$ we get $y = 3(3) + 8 = 17$.

Example 4. Find the equation of the line through the two points $(-2, 6)$ and $(t, t^2 + 2)$, where t is a fixed but arbitrary constant.

Solution. The coordinates of the two given points tell us that $x_1 = -2$, $y_1 = 6$, $x_2 = t$, and $y_2 = t^2 + 2$. Substituting these into the formula for the slope of the line, we get

$$m = \frac{y_2 - y_1}{x_2 - x_1} = \frac{(t^2 + 2) - (6)}{t - (-2)} = \frac{t^2 - 4}{t + 2} = \frac{(t - 2)(t + 2)}{t + 2} = t - 2.$$

We substitute $x_1 = -2$, $y_1 = 6$, and $m = t - 2$ into the general form of the equation of a line, to get

$$y - 6 = (t - 2)(x - (-2))$$

or

$$y = 6 + (t - 2)(x + 2).$$

The slope of this line in the plane depends on the value we assign to the parameter t.

Suppose we are given a parabola in the plane and that P_1 and P_2 denote two points on this parabola. A line through the points P_1 and P_2 is called a secant line for the parabola.

Suppose the plane is a cartesian plane. We almost always work in a cartesian plane. In a cartesian plane all points have coordinates. In a cartesian plane we can describe any parabola by giving its equation. The graph of the equation

$$y = x^2 - 3x - 4$$

is a parabola in the plane. Two points on this parabola are $(0, -4)$ and $(5, 6)$. A line through these two points would be a secant line for the parabola.

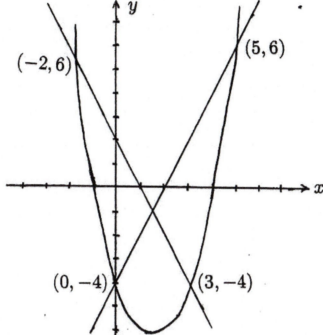

Let us find the equation for this secant line. The slope of the line through the two points $(0, -4)$ and $(5, 6)$ is

$$\text{slope} = m = \frac{y_2 - y_1}{x_2 - x_1} = \frac{6 - (-4)}{5 - 0} = 2.$$

Substituting into $y - y_1 = m(x - x_1)$, the equation of this secant line is

$$y - (-4) = 2(x - 0) \text{ or } y = 2x - 4.$$

Two other points on this same parabola are $(-2, 6)$ and $(3, -4)$. Next let us find the equation of the secant line through these two points. We use $x_1 = 2$, $y_1 = 6$, $x_2 = 3$, and $y_2 = -4$. The slope of the line is

12

$$m = \frac{y_2 - y_1}{x_2 - x_1} = \frac{-4 - (6)}{3 - (-2)} = -2.$$

Substituting into $y - y_1 = m(x - x_1)$ the equation of the secant line is

$$y - (-6) = (-2)(x - 1)$$

$$y = -2x + 2. \qquad \text{\textbullet}$$

Problems

1. Find the equation of the line through the two points $(-2, -4)$ and $(4, 6)$.

2. Find the equation of the line through the two points $(3/4, -2)$ and $(5/3, 5/3)$.

3. Find the equation of the line through the two points $(-2, 4)$ and (t, t^2), where t is a fixed but arbitrary constant.

4. Find the equation of the line through the two points $(3, -5)$ and $(t, 4 - t^2)$ where t is a fixed but arbitrary constant.

5. Consider the parabola whose equation is $y = x^2 - 5x + 4$. Two points on this parabola are $(0, 4)$ and $(3, -2)$. Find the equation of the secant line to this parabola through these two points.

(b) Two other points on this parabola are $(-1, 10)$ and $(5, 4)$. Find the equation of the secant line to this parabola through these two points.

6312 The Tangent Problem

We are going to look at the problem of finding a tangent line to a curve. This problem can be discussed from several points of view, but we will only look at it as a geometry problem. This is a very old and famous geometry problem. The problem is not so famous with today's students because in today's classrooms there is very little study of geometry. In 1950 everyone when starting calculus would have known this problem from analytic geometry and from synthetic geometry.

We will begin by describing the problem. First, we need a curve. Suppose we have an ellipse (flattened circle) or a parabola. If we draw a line intersecting such a curve at two points, then the line is called a secant line for the curve.

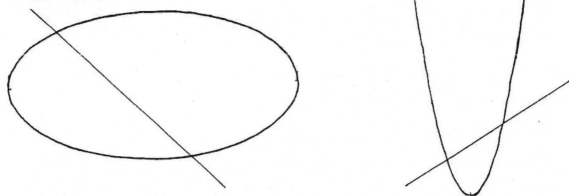

Suppose we fix one point P on the curve and draw several secant lines through the point P and some other points M_1, M_2, M_3, M_4 on the curve.

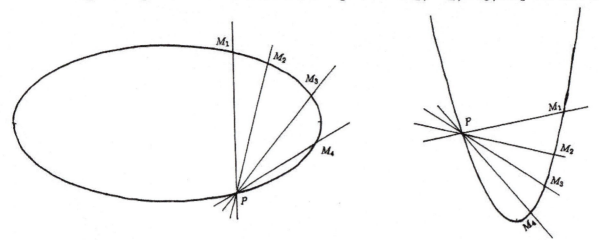

In this figure the points M_1, M_2, M_3, M_4 are located so that each one is closer to the fixed point P than is the previous one. That is, M_3 is closer to P than is M_2. Suppose we think of the point M as a movable point that moves from M_1 to M_2 to M_3 to M_4. As M gets closer to the fixed point P the secant line which passes through P and moving point M keeps changing. What happens to the secant lines when we take the limit as M approaches P? When we take the limit of the secant lines as M approaches P we get the line that intersects the curve twice at the same point P. The limit of such secant lines is called the tangent line. The tangent line crosses the curve twice at the same point. The tangent line just touches (intersects with) the curve at the one point P where the line is tangent to the curve.

The only way that we know how to describe lines and curves is to use coordinate geometry. We describe a curve by giving the equation of the curve. We describe a line by giving the equation of the line. Recall the general equation of a line through the point with coordinates (x_1, y_1) and with slope m the equation is

$$y - y_1 = m(x - x_1).$$

Example 1. The graph of the equation $y = x^2 - 2x$ is a parabola. Find the equation of the secant line to this curve through the points where $x = 1$ and where $x = 4$.

Solution. The y value corresponding to $x = 1$ is $y = 1^2 - 2(1) = -1$. The y value corresponding to $x = 4$ is $y = 4^2 - 2 \cdot 4 = 8$. The secant line is the line through the two points $(1, -1)$ and $(4, 8)$. Now that we know two points on the line we can use the general formula to find the slope of this particular line. The slope is

$$m = \frac{8 - (-1)}{4 - 1} = 3.$$

The equation of the secant line is found by substituting $y_1 = -1$, $x_1 = 1$, and $m = 3$ into the general form as follows:

$$y - (-1) = 3(x - 1)$$
$$y = 3x - 4.$$

Example 2. The graph of the equation $y = x^2 - 3x$ is a parabola. Find the equation of the secant line to this parabola through the point where $x = 1$ and the point where $x = 1 + h$.

Solution. The y value corresponding to $x = 1$, is $y = 1^2 - 3(1) = -2$. The y value corresponding to $x = 1 + h$ is $y = (1 + h)^2 - 3(1 + h) = 1 + 2h + h^2 - 3 - 3h = h^2 - h - 2$. The two coordinates of the points on the secant line are $(1, -2)$ and $(1 + h, h^2 - h - 2)$. We know two points on the secant line so we can substitute into the formula for the slope of a line. The slope of the secant line is

$$m = \frac{(h^2 - h - 2) - (-2)}{1 + h - 1} = \frac{h^2 - h}{h} = h - 1.$$

The secant line is the line through $(1, -2)$ with slope $h - 1$. The equation of the secant line is

$$y - y_1 = m(x - x_1)$$
$$y - (-2) = (h - 1)(x - 1)$$
$$y = -2 + (h - 1)(x - 1).$$

The variables are x and y with h a parameter. The parameter can be assigned **different values** to obtain different secant lines through the point $(1, -2)$.

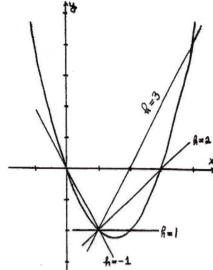

Using different values of h gives different secant lines to the curve all of which pass through the point $(1, -2)$. The smaller the value of h the closer the second point $(1 + h, h^2 - h - 2)$ is to the fixed point $(1, -2)$. If we take the limit of the slope of the secant lines as h approaches zero we get the slope of the tangent line to the parabola at the point $(1, -2)$.

$$\text{Slope of tangent line} = \lim_{h \to 0} (h - 1) = -1.$$

16

The equation of the tangent line to the parabola $y = x^2 - 3x$ at the point $(1, -2)$ is $y + 2 = (-1)(x - 1)$.

Example 3. The graph of the equation $y = x^2 - 6x$ is a parabola. Find the slope of the tangent line to this curve at the point where $x = 2$.

Solution. The problem asks us to find the slope of the tangent line. It does not ask us to find the equation of the tangent line. This is because we want to focus on the problem of finding the slope. Once we know the slope we can easily find the equation of the tangent line if we need to do that.

The y value on the parabola corresponding to $x = 2$ is $y = 2^2 - 6(2) = -8$. We are trying to find the slope of the tangent line to the curve at the point $(2, -8)$. In order to find the slope of the tangent line we begin by looking at the slope of several secant lines. One point on all these secant lines is $(2, -8)$. Let the x coordinate of the other point on the secant line be denoted by $x = t$. For values of t near 2 this is another point on the parabola near the point $(2, -8)$. The y coordinate of the point whose x coordinate is $x = t$ is $y = t^2 - 6t$.

As we change the value of t the point with coordinates $(t, t^2 - 6t)$ moves along the parabola

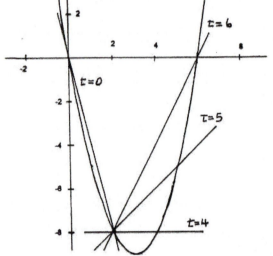

The slope of the secant line through the two points $(2, -8)$ and $(t, t^2 - 6t)$ is

$$\text{slope} = m = \frac{(t^2 - 6t) - (-8)}{(t - 2)} = \frac{(t - 2)(t - 4)}{t - 2} = t - 4.$$

We get the slope of different secant lines by changing the value of t in $t - 4$. All these secant lines pass through $(2, -8)$. The closer the value of t is to

17

2 the closer the point $(t, t^2 - 6t)$ is to the fixed point $(2, -8)$. The slope of the tangent line is obtained by taking the limit of the slope of the secant lines as t approaches 2.

$$\text{Slope of tangent line} = \lim_{t \to 2}(t - 4) = -2.$$

The slope of the tangent line to the curve $y = x^2 - 6x$ at the point $(2, -8)$ is -2. Once we know the slope we can find the equation of the tangent line.

We can even find the slope of the tangent line to this curve at a general point say the point where $x = b$. Note that the value of b is fixed. This is a slightly harder problem. We only consider values of t near b, then the point where $x = t$ is near the point where $x = b$. The y coordinate of the point on the curve whose x coordinate is $x_1 = b$ is $y_1 = b^2 - 6b$. The y coordinate of the point on the curve whose x coordinate is $x_2 = t$ is $y_2 = t^2 - 6t$.

The slope of the secant line through the two points $(x_1, y_1) = (b, b^2 - 6b)$ and $(x_2, y_2) = (t, t^2 - 6t)$ is

$$m = \frac{y_2 - y_1}{x_2 - x_1} = \frac{(t^2 - 6t) - (b^2 - 6b)}{t - b} = \frac{(t^2 - b^2) - 6t + 6b}{t - b}$$
$$= \frac{(t - b)(t + b) - 6(t - b)}{t - b}$$
$$= \frac{(t - b)[t + b - 6]}{t - b} = t + b - 6.$$

The slope of the secant line through the points where $x = b$ and where $x = t$ is given by

$$m = t + b - 6.$$

The limit as $t \to b$ of the secant lines through $x = b$ is the tangent line through the point where $x = b$. The slope of the tangent line through the point where $x = b$ is

$$\text{slope of tangent line} = \lim_{t \to b}(t + b - 6) = 2b - 6.$$

For all values of b the slope of the tangent line to the curve $y = x^2 - 6x$ at the point where $x = b$ is given by slope $= 2b - 6$.

Example 4. Find the slope of the tangent line to the curve whose equation is $y = x^3 - 4x^2 + 6x$ at the general point where $x = b$.

Solution. We think of b as some fixed number. Consider $x = t$ where t is same number close to b. We want to find the slope of the secant line through the fixed point where $x = b$ and through the variable point where $x = t$. The y coordinate of the point on the curve $y = x^3 - 4x^2 + 6x$ corresponding to $x_1 = b$ is $y_1 = b^3 - 4b^2 + 6b$. The y coordinate of the variable point on the curve $y = x^3 - 4x^2 + 6x$ corresponding to $x_2 = t$ is $y_2 = t^3 - 4t^2 + 6t$. We want to look at the secant lines which pass through the two points with coordinates $(b, b^3 - 4b^2 + 6b)$ and $(t, t^3 - 4t^2 + 6t)$. There are many many such lines because while b is fixed t is variable. The slope of the secant line through these two points is

$$m = \frac{y_2 - y_1}{x_2 - x_1} = \frac{(b^3 - 4b^2 + 6b) - (t^3 - 4t^2 + 6t)}{b - t}$$

$$= \frac{b^3 - t^3 - 4(b^2 - t^2) + 6(b - t)}{b - t}.$$

Recall that

$$b^3 - t^3 = (b - t)(b^2 + bt + t^2) \text{ and } b^2 - t^2 = (b - t)(b + t).$$

It follows that

$$m = \frac{(b - t)(b^2 + bt + t^2) - 4(b - t)(b + t) + 6(b - t)}{b - t}$$

$$= \frac{(b - t)[(b^2 + bt + t^2) - 4(b + t) + 6]}{b - t}$$

$$= b^2 + bt + t^2 - 4b - 4t + 6 \text{ for } t \neq b.$$

When we take the limit as t approaches b the secant lines approach the tangent line to the curve $y = x^3 - 4x^2 + 6x$ at the point where $x = b$. Therefore, the slope of the secant lines must approach the slope of the tangent line at the point where $x = b$.

$$\text{slope of the tangent line} = \lim_{t \to b}(b^2 + bt + t^2 - 4b - 4t + 6)$$

$$= b^2 + b^2 + b^2 - 4b - 4b + 6$$

$$= 3b^2 - 8b + 6.$$

The slope of the tangent line to the curve $y = x^3 - 4x^2 + 6x$ at the point where $x = b$ is given by

$$3b^2 - 8b + b.$$

Suppose we think of a general curve $y = f(x)$ and a general point $x = b$. If we continue to work more and more complicated examples such as the one above we will reach the following conclusion.

Theorem. The slope of the tangent line to the curve whose equation is $y = f(x)$ at the point where $x = b$ is given by

$$m = \lim_{t \to b} \frac{f(t) - f(b)}{t - b} = \lim_{x \to b} \frac{f(x) - f(b)}{x - b}.$$

Since it is important that we be able to find the slope of tangent lines it is important that we study the limit

$$\lim_{x \to b} \frac{f(x) - f(b)}{x - b}.$$

Let us look at this limit using a different notation. Suppose we let $x = b + h$, then to say "$\lim_{x \to b}$" is the same as saying "$\lim_{h \to 0}$". Using $x = b + h$ we can write this same limit as

$$\lim_{h \to 0} \frac{f(b + h) - f(b)}{h}.$$

Problems

1. The graph of the equation $y = x^2 - 4x + 5$ is a parabola. Find the equation of the secant line to this parabola through the points where $x = 2$ and $x = 6$.

2. Consider the curve whose equation is $y = x^3 + 2x^2 - 5x$. Find the equation of the secant line to this curve through the points where $x = 1$ and $x = 3$.

3.(a) Consider the parabola whose equation is $y = 2x^2 - 5x$. Find the slope of the secant line of this curve that passes through the points where $x = 2$ and $x = 2 + h$.

(b) What is the slope of the secant line through the two points with x coordinates 2 and 2.5? What is the slope of the secant line through the two points with x coordinates 2 and 2.1?

4. Consider the curve whose equation is $y = x^3 - 5x^2 + 8x$. Find the slope of the secant line to this curve that passes through the points where $x = 3$ and $x = t$.

5.(a) Consider the curve which is the graph of the equation $y = x^3 - 3x^2 + 8x$. Find the slope of the secant line to this curve which passes through the points where $x = 2$ and $x = t$.

(b) Use the slope of the secant lines which you found in part (a) to find the slope of the tangent line to the curve $y = x^3 - 3x^2 + 8x$ at the point where $x = 2$.

6.(a) Consider the curve which is the graph of the equation $y = 3x^2 + 8x$. Find the slope of the secant line to this curve which passes through the points where $x = b$ and $x = t$.

(b) Use the slopes of the secant lines which you found in part (a) to find the slope of the tangent line to the curve $y = 3x^2 + 8x$ at the point where $x = b$.

7. Consider the curve which is the graph of the equation $y = x^3 + 5x^2 - 12x$. Find the slopes of the secant lines that pass through the fixed point where $x = b$ and the variable point where $x = t$.

(b) Use the slopes of the secant lines which you found in part (a) to find the slope of the tangent line to the curve $y = x^3 + 5x^2 - 12x$ at the point where $x = b$.

8. Consider the curve which is the graph of the equation $y = 4/(3x + 5)$. Find the slopes of the secant lines that passes through the fixed point where $x = b$ and the variable point where $x = t$.

(b) Use the slopes of the secant lines which you found in part (a) to find the slope of the tangent line to the curve $y = 4/(3x + 5)$ at the point where $x = b$.

6314 The Velocity Problem

We are going to consider the problem of how to find the velocity of a moving object. Since this is our first try at this problem, let us restrict our attention to the simplest situation. Let us suppose that we have an object that moves only along a straight line. The object could be a car moving along a long straight road. It could be an object falling to the ground from a great height. In order to locate the object we mark off the line along which the object moves like a number line. We can then locate the object by saying that the position of the object is at a certain number. If the object is a car on a straight road, we can write numbers along the road that tell us how many yards the car is from the zero point. The position of the car is indicated by saying that it is at a certain yard marker. Suppose we have a small bug that moves back and forth along a straight line path, then the path is marked by how many inches the bug is on either the negative or positive side of the zero point. In order to work such velocity problems we usually assume that we have a function which gives the position of the object on the line at any time.

In general, suppose an object moves along a straight line according to an equation of motion $s = f(t)$, where s is the displacement of the object from the origin at time t. Negative values of s indicate that the object is on the other side of the zero point. The number $|s|$ is the distance of the object from the zero point. The function $f(t)$ that describes the motion is called the position function of the object.

Example 1. Suppose a ball is thrown toward the ground from a height of 500 feet with an initial velocity of 20 feet per second. Suppose we take ground level as the zero point and up is positive. In this case the position of the ball above ground is given by

$$s = -16t^2 - 20t + 500.$$

When $t = 2$ the corresponding value of s is $s = -16(2)^2 - 20(2) + 500 = 396$. This means that when $t = 2$ seconds the ball is 396 feet above the ground. When $t = 4$ the corresponding value of s is $s = -16(4)^2 - 20(4) + 500 = 164$. This means that when $t = 4$ seconds the ball is 164 feet above the ground. During the time interval from $t = 2$ to $t = 4$ the ball changes from 396 feet to 164 feet above the ground. The net change of position (also called the change in displacement) is

$$164 - 396 = -232 \text{ feet}.$$

The average velocity of an object is defined to be

$$\text{average velocity} = \frac{\text{change in position}}{\text{time elapsed}}.$$

The average velocity of the ball during the time interval $t = 2$ to $t = 4$ seconds is

$$\text{average velocity} = -\frac{232}{2} = -116 \text{ ft/sec}.$$

The fact that the velocity is negative indicates that the velocity is down since up is positive. A displacement of -232 feet also indicates 232 feet downward.

When $t = 5$ the corresponding value of s is $s = -16(5)^2 - 20(5) + 500 = 0$. The ball hits the ground when $t = 5$ seconds.

Let us compute the average velocity over the time period $t = 3$ to $t = u$ seconds. The position of the ball when $t = 3$ seconds is $s = -16(3)^2 - 20(3) + 500 = 296$. The position of the ball when $t = u$ seconds is

$$s = -16u^2 - 20u + 500$$

The change in position over the time period from $t = 3$ to $t = u$ is

$$(-16u^2 - 20u + 500) - (296) = -16u^2 - 20u + 204.$$

Although we like to think that $u > 3$, actually u can be either greater than 3 or less than 3. The average velocity of the ball during the time interval from 3 seconds to u seconds is

$$\text{average velocity} = \frac{-16u^2 - 20u + 204}{u - 3} = -4(4u + 17).$$

Definition. We define the instantaneous velocity (velocity) at time $t = b$ to be the limit as u approaches b of the average velocity over the time interval from $t = b$ to $t = u$.

$$\text{Instantaneous velocity} = \lim_{u \to b} (\text{Average Velocity})$$

The instantaneous velocity of the object when $t = 3$ is

$$\lim_{u \to 3}[-4(4u + 17)] = -116.$$

Example 2. The position (in meters) of a particle moving along a straight line is given by the equation of motion

$$s = 2t^3 + 5t^2 + 8t,$$

where t is measured in seconds. Find the instantaneous velocity when $t = 3$.

Solution. We first find the average velocity of the particle over the time interval from $t = 3$ seconds to $t = u$ seconds. We are using "the time interval from $t = 3$ to $t = u$". In order to make the time interval get smaller and smaller we say $\lim_{u \to 3}$. The value of position s corresponding to $t = 3$ is $s = 2(3)^3 + 5(3)^2 + 8(3) = 123$. The value of position s corresponding to $t = u$ is $s = 2u^3 + 5u^2 + 8u$. The displacement or change in position (final position minus initial position) from $t = 3$ to $t = u$, which is the change in position, is given by

$$(2u^3 + 5u^2 + 8u) - (123).$$

When we divide the change in position by the change in time, we get the average velocity for the time interval:

$$\frac{2u^3 + 5u^2 + 8u - 123}{u - 3}.$$

In order to simplify this we need to factor the numerator. One factor of the numerator is $u - 3$. We divide the numerator by $u - 3$ in order to find the other factor.

$$
\begin{array}{r}
2u^2 +11u +41 \\
u - 3 \overline{)\, 2u^3 +5u^2 +8u - 123} \\
\underline{2u^3 -6u^2 } \\
11u^2 +8u - 123 \\
\underline{11u^2 -33u } \\
41u - 123 \\
\underline{41u - 123}
\end{array}
$$

24

This long division tells us that

$$2u^3 + 5u^2 + 8u - 123 = (u - 3)(2u^2 + 11u + 41)$$

This is the change in position. Note that the change in position can be either positive or negative. The change in time is $u - 3$. The average velocity on the time interval from $t = 3$ to $t = u$ is change in position divided by change in time:

$$\frac{(u - 3)(2u^2 + 11u + 41)}{u - 3} = 2u^2 + 11u + 41.$$

We get the instantaneous velocity when $t = 3$ by taking the limit of the average velocity over the time interval $t = 3$ to $t = u$ as the length of the time interval approaches zero. The length of the time interval is $u - 3$. This means take the limit as u approaches 3.

$$\lim_{u \to 3}(2u^2 + 11u + 41) = 2(3)^2 + 11(3) + 41 = 92$$

The instantaneous velocity is 92 meters/second since we assume that s is measured in meters.

Another problem. Find the average velocity of this particle during the time interval $3 \leq t \leq 5$.

Solution. The average velocity is the change in position (displacement) divided by the change in time. The initial position when $t = 3$ is $y = 2(3)^3 + 5(3)^2 + 8(3) = 123$. The final position when $t = 5$ is $y = 2(5)^3 + 5(5)^2 + 8(5) = 415$. The change in position is the final position minus the initial position. This is $415 - 123 = 292$. The change in time is $5 - 3 = 2$. Dividing the change in position by the change in time, we get

$$\text{Average velocity } = \frac{292}{2} = 146.$$

Example 3. The position (in feet) of a particle moving in a straight line is given by the equation of motion

$$s = 3t^2 + 8t + 12$$

25

where t is measured in seconds. Find the instantaneous velocity when $t = b$.

Solution. The time $t = b$ is fixed. Let $t = u$ seconds denote a general time near $t = b$. The position s corresponding to $t = b$ is $s = 3(b)^2 + 8(b) + 12$. The position s corresponding to $t = u$ seconds is $s = 3u^2 + 8u + 12$ feet. First, we need to find the change in position of the particle during the time interval from when $t = b$ seconds to the time when $t = u$ seconds. This change in position is given by the final position minus the initial position.

$$
\begin{aligned}
(3u^2 + 8u + 12) - (3b^2 + 8b + 12) &= 3(u^2 - b^2) + 8(u - b) \\
&= 3(u + b)(u - b) + 8(u - b) \\
&= [3(u + b) + 8](u - b)
\end{aligned}
$$

The amount of elapsed time from when $t = b$ to when $t = u$ is $u - b$. Note that $u - b$ could be negative, that is, u could be less than b. We divide the change in position by the change in time to get the average velocity:

$$
\text{average velocity} = \frac{(u - b)[3(u + b) + 8]}{u - b} = 3u + 3b + 8.
$$

This is the average velocity of the particle during the time interval from $t = b$ to $t = u$. In order to find the instantaneous velocity when $t = b$ we look at shorter and shorter time intervals. We take the limit as u approaches b of the average velocity and get

$$
\lim_{u \to b} (3u + 3b + 8) = 6b + 8.
$$

The instantaneous velocity at time $t = b$ is $6b + 8$ feet per second.

As we look at more and more examples of finding instantaneous velocity we see a pattern. The pattern can be described in the general case as follows. Suppose an object moves along a straight line according to the equation of motion $s = f(t)$, where s is the position (directed distance) of the object from the origin at time t. In the time interval from $t = b$ to $t = u$ the change of position (displacement) is $f(u) - f(b)$. The change in time is $u - b$. The average velocity is the change in position divided by change in time:

$$
\text{average velocity} = \frac{f(u) - f(b)}{u - b}.
$$

We get the instantaneous velocity at the time $t = b$ by taking the limit as the length of the time interval approaches zero, that is, the limit as u approaches b.

$$\text{instantaneous velocity} = \lim_{u \to b} \frac{f(u) - f(b)}{u - b}.$$

This limit is the instantaneous velocity at $t = b$, that is, $v(b)$.

Problems

1. Suppose a ball is thrown upward with an initial velocity of 300 ft/sec on a planet far away, then the height s of the ball at any time t is given by $s = 300t - 40t^2$.

(a) Find the average velocity of the ball during the time interval from $t = 2$ to $t = 5$.

(b) Find the average velocity of the ball during the time interval from $t = 4$ to $t = 7$.

(c) Find the average velocity of the ball during the time interval from $t = 3$ to $t = u$.

(d) Use the result of part (c) to find the instantaneous velocity of the ball at time $t = 3$.

2. The position in feet of a particle moving in a straight line is given by the equation of motion

$$s = 3t^2 - 4t + 12$$

where t is measured in seconds.

(a) Find the average velocity for this particle during the time interval from $t = b$ to $t = u$.

(b) Use the result from part (a) to find the instantaneous velocity of the particle at time $t = b$.

3. The position in feet of a particle moving in a straight line is given by the equation of motion

$$s = t^3 - 6t^2 + 10t$$

where t is measured in seconds.

(a) Find the average velocity for this particle during the time interval from $t = 2$ to $t = u$.

(b) Use the result from part (a) to find the instantaneous velocity of the particle at time $t = 2$.

(c) Find the average velocity for this particle during the time interval from $t = b$ to $t = u$.

(d) Use the result of part (c) to find the instantaneous velocity of the particle at time $t = b$.

6315 Definition of Derivative

We have seen that the slope of the tangent line to a curve with equation $y = f(x)$ at the point where $x = b$ is given by

$$m = \lim_{t \to b} \frac{f(t) - f(b)}{t - b} = \lim_{x \to b} \frac{f(x) - f(b)}{x - b}.$$

Suppose an object is moving along a straight line with position function $s = f(t)$. The instantaneous velocity $v(b)$ of the object at time $t = b$ is given by

$$v(b) = \lim_{t \to b} \frac{f(t) - f(b)}{t - b} = \lim_{u \to b} \frac{f(u) - f(b)}{u - b}.$$

Note that both these limits are the same. In fact in order to calculate the rates of change in many other situations we find that we must also evaluate this very same limit. This is clearly an important limit. Also note that this is a two sided limit. Since this limit occurs so widely it is given a special name.

Definition of Derivative. The derivative of the function $f(x)$ at the fixed number $x = b$, denoted by $f'(b)$, is

$$f'(b) = \lim_{x \to b} \frac{f(x) - f(b)}{x - b},$$

whenever this limit exists. If we define h by $x = b + h$, then this same limit can be written as

$$f'(b) = \lim_{h \to 0} \frac{f(b + h) - f(b)}{h}.$$

Definition. When working with derivatives the fraction

$$\frac{f(x) - f(b)}{x - b}$$

is called the difference quotient.

When we discussed the problem of finding a tangent line the difference quotient was the slope of the secant line. When we discussed velocity the difference quotient was the average velocity.

Example 1. Given $f(x) = 5x^2 + 8x$, find the derivative of $f(x)$ when $x = 2$. First, find the difference quotient.

Solution. When $x = 2$, we have $f(2) = 5 \cdot 4 + 8 \cdot 2 = 36$. First, find the difference quotient which is

$$\frac{f(x) - f(2)}{x - 2} = \frac{(5x^2 + 8x) - 36}{x - 2}.$$

After factoring we can reduce this fraction as follows:

$$\frac{5x^2 + 8x - 36}{x - 2} = \frac{(x - 2)(5x + 18)}{x - 2} = 5x + 18 \text{ for } x \neq 2.$$

It follows that

$$\lim_{x \to 2} \frac{5x^2 + 8x - 36}{x - 2} = \lim_{x \to 2} (5x + 18).$$

Taking the limit as x approaches 2, we get the derivative at the point where $x = 2$:

$$f'(2) = \lim_{x \to 2} (5x + 18) = 28.$$

Example 2. Find the derivative of $f(x) = 5x^2 + 8x$ at the fixed point $x = b$. First, find the difference quotient.

Solution. When $x = b$ the value of the function is $f(b) = 5b^2 + 8b$. The difference quotient is

$$\frac{f(x) - f(b)}{x - b} = \frac{(5x^2 + 8x) - (5b^2 + 8b)}{x - b}.$$

Factoring, we can reduce this fraction as follows:

$$\frac{5x^2 + 8x - 5b^2 - 8b}{x - b} = \frac{5(x^2 - b^2) + 8(x - b)}{x - b}$$
$$= \frac{5(x - b)(x + b) + 8(x - b)}{x - b}$$
$$= \frac{(x - b)[5(x + b) + 8]}{x - b}$$
$$= 5(x + b) + 8.$$

It follows that

$$\lim_{x \to b} \frac{5x^2 + 8x - 5b^2 - 3b}{x - b} = \lim_{x \to b} [5(x + b) + 8].$$

Taking the limit as x approaches b, we get the derivative of $f(x)$ for any constant b:

$$f'(b) = \lim_{x \to b} 5(x + b) + 8 = 10b + 8.$$

When $b = 4$ the derivative is $f'(4) = 10(4) + 8 = 48$. We can substitute any number in for b and find the derivative of $f(x) = 5x^2 + 8x$ for that number. We would probably write this as $f'(x) = 10x + 8$.

Example 3. Given $f(x) = 4x^3 + 5x^2 + 6x$, find the derivative when $x = 2$. First, find the difference quotient.

Solution. First, note that the value of $f(x)$ when $x = 2$ is given by $f(2) = 4 \cdot 2^3 + 5 \cdot 2^2 + 6 \cdot 2 = 64$. The difference quotient is

$$\frac{f(x) - f(2)}{x - 2} = \frac{4x^3 + 5x^2 + 6x - 64}{x - 2}.$$

Using long division

$$
\begin{array}{r}
4x^2 \ +13x \ +32 \\
x - 2 \ \overline{\smash{\big)}\ 4x^3 \ +5x^2 \ +6x - 64} \\
\underline{4x^3 \ -8x^2} \\
13x^2 \ +6x - 64 \\
\underline{13x^2 \ -26x} \\
32x - 64 \\
\underline{32x - 64} \\
\end{array}
$$

This long division tells us that

$$4x^3 + 5x^2 + 6x - 64 = (x - 2)(4x^2 + 13x + 32).$$

We can now write the difference quotient as

$$\frac{(x - 2)(4x^2 + 13x + 32)}{x - 2} = 4x^2 + 13x + 32, \text{ for } x \neq 2.$$

31

Taking the limit as x approaches 2, we get the derivative

$$f'(2) = \lim_{x \to 2} (4x^2 + 13x + 32) = 74.$$

Example 4. Given $f(x) = 4x^3 + 5x^2 + 6x$, find the derivative when $x = b$.

Solution. The value of $f(x)$ when $x = b$ is $f(b) = 4b^3 + 5b^2 + 6b$. The difference quotient is

$$\frac{f(x) - f(b)}{x - b} = \frac{(4x^3 + 5x^2 + 6x) - (4b^3 + 5b^2 + 6b)}{x - b}.$$

This can be rewritten as

$$\frac{f(x) - f(b)}{x - b} = \frac{4(x^3 - b^3) + 5(x^2 - b^2) + 6(x - b)}{x - b}.$$

Recall that

$$x^2 - b^2 = (x - b)(x + b)$$
$$x^3 - b^3 = (x - b)(x^2 + bx + b^2).$$

Using these expressions the difference quotient fraction can be rewritten as

$$\frac{4(x - b)(x^2 + bx + b^2) + 5(x - b)(x + b) + 6(x - b)}{x - b}$$
$$= \frac{(x - b)[4(x^2 + bx + b^2) + 5(x + b) + 6]}{x - b}$$
$$= 4(x^2 + bx + b^2) + 5(x + b) + 6, \text{ for } x \neq b.$$

The derivative at $x = b$ is found by taking the limit as x approaches b.

$$f'(b) = \lim_{x \to b} 4(x^2 + bx + b^2) + 5(x + b) + 6$$
$$= 4(b^2 + b^2 + b^2) + 5(b + b) + 6$$
$$= 12b^2 + 10b + 6.$$

Since "b" can be any number, we often write x instead of b, that is, we write $f'(x) = 12x^2 + 10x + 6$ instead of $f'(b) = 12b^2 + 10b + 6$. This formula gives the derivative for any value of b. For instance, the derivative at $b = 5$ is

$$f'(5) = 12(5)^2 + 10(5) + 6 = 356.$$

Example 5. Let $f(x) = \dfrac{1}{3x+5}$, find $f'(b)$ where b is any constant. Note that the function $f(x)$ is not defined for $x = -5/3$. This means $b \neq -5/3$.

Solution. The difference quotient is

$$\frac{f(x) - f(b)}{x - b} = \frac{\frac{1}{3x+5} - \frac{1}{3b+5}}{x - b}.$$

We need to simplify this complex fraction. Lets at first consider the numerator of the fraction only:

$$\frac{1}{3x+5} - \frac{1}{3b+5} = \frac{(3b+5)}{(3x+5)(3b+5)} - \frac{3x+5}{(3x+5)(3b+5)}$$
$$= \frac{3b+5-3x-5}{(3x+5)(3b+5)} = \frac{3b-3x}{(3x+5)(3b+5)}.$$

Using this expression for the numerator, we can then write the fraction as

$$\frac{\frac{-3(x-b)}{(3x+5)(3b+5)}}{(x-b)}.$$

In order to simplify a complex fraction we take the numerator and divide by the denominator

$$\frac{-3(x-b)}{(3x+5)(3b+5)} \div (x-b).$$

In order to divide we invert and multiply as follows:

$$\frac{-3(x-b)}{(3x+5)(3b+5)} \cdot \frac{1}{x-b} = \frac{-3}{(3x+5)(3b+5)}.$$

Thus we see that the difference quotient is given by

$$\frac{f(x) - f(b)}{x - b} = \frac{-3}{(3x+5)(3b+5)}, \quad \text{when } x \neq b.$$

In order to find the derivative at the point $x = b$ we take the limit of the difference quotient as x approaches b.

$$f'(b) = \lim_{x \to b} \frac{-3}{(3x+5)(3b+5)} = \frac{-3}{(3b+5)(3b+5)} = \frac{-3}{(3b+5)^2}$$

The derivative of $f(x)$ at $x = b$ is defined as

$$f'(b) = \lim_{x \to b} \frac{f(x) - f(b)}{x - b}.$$

This is a two sided limit. In stating this limit it is assumed that if $\epsilon > 0$ and ϵ is small enough, then $f(x)$ is defined for all x such that $b - \epsilon < x < b + \epsilon$. This says that in order to find $f'(b)$ the function $f(x)$ must be defined for values of x or both sides of b. Suppose $f(x)$ is only defined for $c \le x \le d$, then the derivative as defined by this limit does not define $f'(c)$ or $f'(d)$. The expressions $f'(c)$ and $f'(d)$ when $f(x)$ is only defined for $c \le x \le d$ will be defined later.

Exercises

1. If $f(x) = x^3 - 3x^2 + 6x$, find $f'(2)$ using the definition of derivative. Use $f'(2)$ to find an equation of the tangent line to the curve $y = x^2 - 3x^2 + 6x$ at the point on the curve where $x = 2$.

2. If $f(x) = \dfrac{x}{3x-2}$, find $f'(2)$ using the definition of derivative. Explain in words what you are doing at each step.

3. Each of the following limits represents the derivative $f'(b)$ for some function f and some number b. State both the function f and the constant b in each case.

(a) $\lim_{x \to 2} \dfrac{x^4 - 16}{x - 2}$

(b) $\lim_{x \to 2} \dfrac{2x^2 + 7x - 22}{x - 2}$

(c) $\lim_{x \to 4} \dfrac{\sqrt{x^2 + 9} - 5}{x - 4}$

(d) $\lim_{x \to 3} \dfrac{\frac{1}{2x+1} - \frac{1}{7}}{x - 3}$

4. If $f(x) = 3x^3 + 10x$ find $f'(b)$ using the definition of derivative. Show all steps and explain in words what you are doing.

5. If $f(x) = x^3 - 6x^2 + 10x$, find $f'(b)$ using the definition of derivative. Find all values of b such that $f'(b) = 1$.

6. If $f(x) = \dfrac{x}{4x + 3}$, find $f'(b)$ where b is a constant with $b \neq -3/4$ using the definition of derivative.

7. If $f(x) = \sqrt{3x + 1}$, find $f'(5)$ using the definition of derivative. Explain in words what you are doing at each step.

8. First show that

$$\frac{x^{3/2} - b^{3/2}}{x - b} = \frac{(x^{1/2} - b^{1/2})(x + x^{1/2}b^{1/2} + b)}{(x^{1/2} - b^{1/2})(x^{1/2} + b^{1/2})}$$

with this fact in mind, let $f(x) = x^{3/2}$ and find $f'(b)$.

9. Find the limit

$$\lim_{t \to x} \frac{\frac{1}{x^2} - \frac{1}{t^2}}{x - t}.$$

6316 The Power Formula

Example 1. Let $f(x) = x^3$, find $f'(b)$.

Solution. The difference quotient is

$$\frac{x^3 - b^3}{x - b}.$$

Factoring and reducing we get

$$\frac{x^3 - b^3}{x - b} = \frac{(x - b)(x^2 + bx + b^2)}{(x - b)} = x^2 + bx + b^2.$$

The derivative is

$$f'(b) = \lim_{x \to b}(x^2 + bx + b^2) = 3b^2.$$

Since "b" is an arbitrary constant, we can replace "b" with any number. For example,

$$f'(4) = 3(4^2) = 48.$$

Actually, when writing the derivative function, we usually use the same letter for the independent variable in the derivative function as we did in the function itself. This means that for this problem we write $f'(x) = 3x^2$ for the derivative. We understand that we can replace x in $f'(x)$ with any number, say 5, and get the value of the derivative of $f(x)$ for that value of x, say $f'(5) = 75$.

Example 2. Suppose n is a positive integer, find the derivative of $f(x) = x^n$.

Solution. This problem is very similar to Example 1 except that the algebra is more complicated. We skip the algebra. The result is

$$f'(x) = nx^{n-1}.$$

In general we sometimes use the notation $\frac{d}{dx}[f(x)]$ instead of $f'(x)$. Using this notation we would write this derivative formula as

$$\frac{d}{dx}[x^n] = nx^{n-1}.$$

36

In the above discussion of powers when we wrote $f(x) = x^m$ we assumed that m was a positive integer. When m takes on different positive integer values we get, for example, x^3, x^8, and x^{15}. From our previous work in algebra we know that we also want to talk about powers that are fractions. Let us review some of the basic ideas of fractional powers.

We want to review the definition of fractional powers. Let us start with the simplest fractional power, $f(x) = x^{1/2}$, which is the same as $f(x) = \sqrt{x}$. By definition $\sqrt{x}$ is a number whose square is x, that is, $\sqrt{x}$ is a number such that $(\sqrt{x})^2 = x$. This says $\sqrt{25}$ is a number such that $(\sqrt{25})^2 = 25$. It is easy to decide that $\sqrt{25} = 5$ because $5^2 = 25$. Also $\sqrt{12}$ is a number such that $(\sqrt{12})^2 = 12$. It is not possible to express $\sqrt{12}$ in the simple way that we could express $\sqrt{25}$. When discussing the expression $\sqrt{25}$ there is a difficulty since $(-5)^2 = 25$. There are two real numbers that are $\sqrt{25}$. Since each single symbol should have a unique meaning, we need to reach an understanding. When we write "$\sqrt{25}$" we mean 5. We use $-\sqrt{25}$ to mean -5. In general, if $x > 0$, then $x^{1/2} = \sqrt{x}$ is the positive number such that $(\sqrt{x})^2 = x$. If we want to talk about the negative number which is the square root of x, we write $(-x^{1/2}) = -\sqrt{x}$.

There is another difficulty connected with $x^{1/2}$. Recall that $\sqrt{-9} = 3i$. If x is negative, none of the square roots of x are real numbers. In calculus we are dealing only with real numbers. Therefore, in calculus we say that $\sqrt{x} = x^{1/2}$ is not defined for $x < 0$. The function $f(x) = x^{1/2}$ is only defined for $x \geq 0$ and has a value such that $x^{1/2} \geq 0$.

Next, let us consider the function $g(x) = x^{1/3}$. Recall that $x^{1/3} = \sqrt[3]{x}$ is defined to be a number whose cube is x, that is, $\sqrt[3]{x}$ is a number such that $(\sqrt[3]{x})^3 = x$. We know that $\sqrt[3]{64} = 4$ because $4^3 = 64$. Is it possible to make sense out of the expression $x^{1/3}$ when x is negative? Yes, we just take a negative value. For example, $\sqrt[3]{-27} = -3$ because $(-3)^3 = -27$. There is a real value for $x^{1/3} = \sqrt[3]{x}$ for all values of x. In fact, if we deal in complex numbers there are three values of $\sqrt[3]{x}$. However, two of these values are complex numbers and so we ignore them. In summary, there is a unique real number $\sqrt[3]{x}$ such that $(\sqrt[3]{x})^3 = x$. The function $g(x) = x^{1/3}$ is defined for all x.

Consider the function $f(x) = x^{1/4}$. The number $x^{1/4} = \sqrt[4]{x}$ is defined to be the number such that $(\sqrt[4]{x})^4 = x$. For example, $\sqrt[4]{625} = 5$ because $5^4 = 625$. On the other hand $\sqrt[4]{625} = -5$ because $(-5)^4 = 625$. We also see that $\sqrt[4]{x}$ is the same as the square root of the square root of x. The function

$x^{1/4}$ has essentially the same properties as $x^{1/2}$. The function $x^{1/4}$ is only defined for $x \geq 0$. We do not write $(-256)^{1/4}$. There are four complex numbers $\sqrt[4]{x}$ such that $(\sqrt[4]{x})^4 = x$. However, when we write $x^{1/4} = \sqrt[4]{x}$ we mean the positive value, that is, $x^{1/4} \geq 0$.

In general, the function $x^{1/n}$ where n is an even positive integer is similiar to the function $x^{1/2}$ and the function $x^{1/4}$. The function $x^{1/n}$, n even, is defined only when $x \geq 0$. We choose the value of $x^{1/n}$, n even, such that $x^{1/n} \geq 0$.

Next, consider the function $x^{1/5}$. The number $x^{1/5} = \sqrt[5]{x}$ is defined to be a number such that $(\sqrt[5]{x})^5 = x$. The function $x^{1/5}$ is very similiar to $x^{1/3}$. Even when x is negative there is a real number which is $x^{1/5} = \sqrt[5]{x}$. The function $x^{1/5}$ is defined for all values of x. In general, the function $x^{1/n}$ where n is an odd positive whole number is similiar to $x^{1/3}$. The function $x^{1/n}$, n odd, is defined for all values of x.

Next, we move on to discuss powers where the numerator of the fraction is a positive integer larger than 1. We want to discuss how to define the function $x^{m/n}$ where m/n is a positive fraction. All functions with fractional powers, $x^{m/n}$, are defined in terms of $x^{1/n}$.

Definition. $x^{m/n} = (x^{1/n})^m$.

This says that in order to find $x^{m/n}$ we first find the number $x^{1/n}$ and then raise this number to the mth power.

Example 3. Find $(64)^{5/3}$. First, we find $(64)^{1/3}$ which is 4 since $4^3 = 64$. Next, we find 4^5 which is 1024. Therefore, $(64)^{5/3} = 1024$.

Note that $x^{m/n}$ is defined for all x if n is odd. The function $x^{m/n}$ is only defined for $x \geq 0$ when n is even. Also $x^{m/n} \geq 0$ when n is even.

Now let us return to the problem: what is the derivative of the function $f(x) = x^{m/n}$? According to the definition of derivative the derivative at $x = b$ is given by the limit

$$f'(b) = \lim_{x \to b} \frac{x^{m/n} - b^{m/n}}{x - b}.$$

We need a lot of algebra to simplify this fraction. If we wanted to wait, we could show that the differentiation rule given below is correct without a lot of algebra by using facts we learn when we study the chain rule and implicit differentiation. However, we want the rule now and so we just give the result here.

$$f'(b) = \lim_{x \to b} \frac{x^{m/n} - b^{m/n}}{x - b} = \frac{m}{n} b^{(m/n)-1}.$$

This is the same formula for finding the derivative of x to a power that was found above. We usually write this as

$$\frac{d}{dx}(x^{m/n}) = (m/n)x^{(m/n)-1}.$$

Next, we will consider how to differentiate when the power on x is negative. Negative powers are defined by the relation

$$x^{-m/n} = \frac{1}{x^{m/n}}.$$

Let $f(x) = x^{-m/n}$ where m and n are positive whole numbers, that is, $(-m/n)$ is a negative fraction. We can find the derivative of

$$f(x) = x^{-m/n} = \frac{1}{x^{m/n}},$$

using the quotient rule which we will learn in a few days. The result is

$$\frac{d}{dx}(x^{-m/n}) = (-m/n)x^{(-m/n)-1}.$$

This is the same power rule.

We need to briefly consider powers that are not fractions. What do we mean by $5^{\sqrt{2}}, 8^{\pi}$? Numbers such as $5^{\sqrt{2}}, 8^{\pi}$, and $6^{\sqrt{5}}$ are defined using limits of a kind that we do not wish to study at this time. Numbers such as $5^{\sqrt{2}}, 8^{\pi}$, and $6^{\sqrt{5}}$ can be approximated for us by our calculator. We will assume that expressions of the form x^a, where a is not a fraction, are defined and that x^a obeys all the usual rules for powers. This includes the power formula for derivatives,

$$\frac{d}{dx}(x^a) = ax^{a-1}.$$

We are now able to conclude that the following theorem is true.

Theorem 1. The Power Rule. Suppose n is either a positive or negative whole number or fraction, or even irrational number, and $f(x) = x^n$, then the derivative is

$$f'(x) = nx^{n-1} \text{ or } \frac{d}{dx}[x^n] = nx^{n-1}.$$

Example 4. Let $f(x) = x^{15}$, find $f'(x)$.

Solution. For this example $n = 15$ and so the Power Rule says

$$f'(x) = 15x^{14} \text{ or } \frac{d}{dx}[x^{15}] = 15x^{14}.$$

Example 5. Let $f(x) = x^{-7/3}$, find $f'(x)$.

Solution. For this example $n = -7/3$ and $n - 1 = -7/3 - 1 = -10/3$. The Power Rule says
$$f'(x) = -(7/3)x^{-(10/3)}.$$

Theorem 2. If $f(x)$ is a function of x and c is a constant, then

$$\frac{d}{dx}[cf(x)] = c\frac{d}{dx}[f(x)].$$

Theorem 3. If $f(x)$ and $g(x)$ are two functions of x, then

$$\frac{d}{dx}[f(x) + g(x)] = \frac{d}{dx}[f(x)] + \frac{d}{dx}[g(x)] = f'(x) + g'(x).$$

These two theorems are proved using the corresponding limit laws.

Example 6. Let $f(x) = 15x^8 + 7x^4$, find the derivative $f'(x)$.

Solution. First, we use Theorem 3 as follows:

$$\frac{d}{dx}[15x^8 + 7x^4] = \frac{d}{dx}[15x^8] + \frac{d}{dx}[7x^4].$$

Theorem 3 says that the derivative of a sum is the sum of the derivatives. Next, we use Theorem 2 to factor out the constants.

$$\frac{d}{dx}[15x^8] + \frac{d}{dx}[7x^4] = 15\frac{d}{dx}[x^8] + 7\frac{d}{dx}[x^4].$$

Finally, we use Power Rule to differentiate powers

$$15\frac{d}{dx}[x^8] + 7\frac{d}{dx}[x^4] = 15(8x^7) + 7(4x^3) = 120x^7 + 28x^3.$$

We have that

$$f'(x) = \frac{d}{dx}[15x^8 + 7x^4] = 120x^7 + 28x^3.$$

Example 7. Let $g(x) = 10x^{9/2} - 8x^{-3} - 6x^{-2/3}$, find $g'(x)$.

Solution. We first use Theorem 3.

$$\frac{d}{dx}[10x^{9/2} - 8x^{-3} - 6x^{-2/3}] = \frac{d}{dx}[10x^{9/2}] + \frac{d}{dx}[-8x^{-3}] + \frac{d}{dx}[-6x^{-2/3}].$$

Second, we use Theorem 2.

$$\frac{d}{dx}[10x^{9/2}] + \frac{d}{dx}[-8x^{-3}] + \frac{d}{dx}[-6x^{-2/3}] =$$
$$10\frac{d}{dx}[x^{9/2}] - 8\frac{d}{dx}[x^{-3}] - 6\frac{d}{dx}[x^{-2/3}].$$

Third, we use Power Rule,

$$10\frac{d}{dx}[x^{9/2}] - 8\frac{d}{dx}[x^{-3}] - 6\frac{d}{dx}[x^{-2/3}]$$
$$= 10(9/2)x^{7/2} - 8(-3)x^{-4} - 6(-2/3)x^{-5/3}$$
$$= 45x^{7/2} + 24x^{-4} + 4x^{-5/3}.$$

Therefore,

$$g'(x) = 45x^{7/2} + 24x^{-4} + 4x^{-5/3}.$$

Example 8. Let $f(x) = \dfrac{3x^4 - 7x}{x^{3/2}}$, find the derivative $f'(x)$.

Solution. We first simplify $f(x)$ as follows:

$$f(x) = \frac{3x^4 - 7x}{x^{3/2}} = \frac{3x^4}{x^{3/2}} - \frac{7x}{x^{3/2}} = 3x^{5/2} - 7x^{-1/2}.$$

Using Theorem 3

$$\frac{d}{dx}[3x^{5/2} - 7x^{-1/2}] = \frac{d}{dx}[3x^{5/2}] + \frac{d}{dx}[-7x^{-1/2}].$$

Using Theorem 2:

$$\frac{d}{dx}(3x^{5/2}) + \frac{d}{dx}(-7x^{-1/2}) = 3\frac{d}{dx}(x^{5/2}) - 7\frac{d}{dx}(x^{-1/2}).$$

Using Power Rule:

$$3\frac{d}{dx}(x^{5/2}) - 7\frac{d}{dx}(x^{-1/2}) = 3(5/2)x^{3/2} - 7(-1/2)x^{-3/2}$$
$$= (15/2)x^{3/2} + (7/2)x^{-(3/2)}.$$

Therefore, $f'(x) = (15/2)x^{3/2} + (7/2)x^{-(3/2)}$.

We would now like to discuss a different kind of function involving powers. For this function the constant is the base and x is the power. We can now discuss the function $f(x) = b^x$, where b is a fixed constant. However, at the moment we only wish to discuss such functions in the case $b = e$. We wish to consider the function $f(x) = e^x$. The constant e, which is the base in natural logarithms, can be defined in several ways.

First, using the kind of limits we have already studied the number e can be defined as $e = \lim\limits_{x \to 0} (1 + x)^{1/x}$.

Second, using sequences, which we have not yet studied, we can say $e = \lim\limits_{n \to \infty} (1 + \dfrac{1}{n})^n$.

Third, we can define e by saying that e is the unique number such that $\lim\limits_{h \to 0} \dfrac{e^h - 1}{h} = 1$.

We should really prove that all three of these ways of defining e give the same number. Showing that the first and third statements are equivalent is hard. The third definition is the one that we will use. We need to know the third method for defining e in order to find the derivative of e^x.

Theorem 4. $\dfrac{d}{dx}(e^x) = e^x$.

Proof. We compute the derivative of $f(x) = e^x$ by using the definition of derivative. As stated earlier we assume that e^x obeys the usual rules for powers. We will use the alternate form of the definition that contains the increment h.

$$f'(x) = \lim_{h \to 0} \frac{f(x+h) - f(x)}{h} = \lim_{h \to 0} \frac{e^{x+h} - e^x}{h}$$

$$= \lim_{h \to 0} \frac{e^x(e^h - 1)}{h} = e^x \lim_{h \to 0} \frac{e^h - 1}{h} = e^x(1) = e^x.$$

Theorem 1, 2, and 3 enable us to differentiate polynomial type functions. However, it is easier to differentiate some of these functions when they are expressed as a product: In order to differentiate products, we need the following rule.

Theorem 5. The Product Rule. If $f(x)$ and $g(x)$ are both differentiable, then

$$\frac{d}{dx}[f(x)g(x)] = f(x)\frac{d}{dx}[g(x)] + g(x)\frac{d}{dx}[f(x)].$$

This theorem says: "The derivative of the product of two functions is given by the first function times the derivative of the second function plus the second function times the derivative of the first function". In order to give a detailed proof of the product rule for derivatives we need to use the product rule for limits. We did not give a formal statement of the product rule for limits in our earlier study of limits.

Example 9. Let $F(x) = (5x^3 + 8x)(x^{-2} + 3x^2)$, find $F'(x)$. We often write this problem using the following alternate notation. Let $y = (5x^3 + 8x)(x^{-2} + 3x^2)$, find $\frac{dy}{dx}$.

43

Solution. We think of $F(x)$ as the product $f(x)g(x)$ where $f(x) = 5x^3 + 8x$ and $g(x) = x^{-2} + 3x^2$. The product rule says

$$\frac{d}{dx}[(5x^3 + 8x)(x^{-2} + 3x^2)] =$$

$$(5x^3 + 8x)\frac{d}{dx}(x^{-2} + 3x^2) + (x^{-2} + 3x^2)\frac{d}{dx}(5x^3 + 8x).$$

Using Theorems 1,2, and 3 we get

$$\frac{d}{dx}(x^{-2} + 3x^2) = \frac{d}{dx}(x^{-2}) + 3\frac{d}{dx}(x^2) = -2x^{-3} + 6x$$

$$\frac{d}{dx}(5x^3 + 8x) = 5\frac{d}{dx}(x^3) + 8\frac{d}{dx}(x) = 15x^2 + 8(1).$$

Therefore,

$$\frac{d}{dx}[(5x^3 + 8x)(x^{-2} + 3x^2)] =$$

$$(5x^3 + 8x)(-2x^{-3} + 6x) + (x^{-2} + 3x^2)(15x^2 + 8).$$

If we need to, we can multiply these factors and simplify to get:

$$75x^4 + 72x^2 + 5 - 8x^{-2}.$$

Example 10. Let $G(x) = e^x(5x^4 + 3x^{-2})$, find $G'(x)$.

Solution. This is a product. The product rule says

$$\frac{d}{dx}[e^x(5x^4 + 3x^{-2})] = e^x\frac{d}{dx}(5x^4 + 3x^{-2}) + (5x^4 + 3x^{-2})\frac{d}{dx}(e^x)$$
$$= e^x(20x^3 - 6x^{-3}) + (5x^4 + 3x^{-2})e^x$$
$$= e^x(5x^4 + 20x^3 + 3x^{-2} - 6x^{-3}).$$

Example 11. This example uses the product rule keeping the symbolic notation. Suppose $f(x) = x^{5/2}g(x)$ and that $g(4) = 5$, $g'(4) = 6$, $g(9) = -2$ and $g'(9) = -1$, find $f(4)$, $f'(4)$, $f(9)$, and $f'(9)$.

Solution. Replacing x with 4, we get

$$f(4) = 4^{5/2}g(4) = 32g(4) = 32(5) = 160$$

Replacing x with 9, we get

$$f(9) = 9^{5/2}g(9) = 243g(9) = 243(-2) = -486.$$

Using the product rule for differentiating, we find

$$f'(x) = x^{5/2}g'(x) + (5/2)x^{3/2}g(x).$$

Substituting $x = 4$ into this, we get

$$\begin{aligned} f'(4) &= 4^{5/2}g'(4) + (5/2)(4)^{3/2}g(4) \\ &= 32(6) + (5/2)(8)(5) = 292. \end{aligned}$$

Substituting $x = 9$ into this, we get

$$\begin{aligned} f'(9) &= 9^{5/2}g'(9) + (5/2)(9)^{3/2}g(9) \\ &= 243(-1) + (5/2)(27)(-2) = -378. \end{aligned}$$

Problems

1. Find the derivatives of each of the following functions.

a. $f(x) = 3x^{5/2} + 9x^{-2/3}$.

b. $f(x) = 5x^3 - 8/x^2$.

c. $f(x) = 10x + 7x^{2/5} - 3x^{-5/2}$.

d. $f(x) = \dfrac{x^{5/2} + 4x^2 + x^{-3}}{x^2}$.

e. $f(x) = \dfrac{8x^{3/2} - 3x^{5/2} + 7x^{-1/2}}{x^{3/2}}$.

2. Let $f(x) = 2x^2 + 108x^{-1}$ for $x \neq 0$. Find all values of x such that $f'(x) = 0$.

3. Find the derivative of each of the following functions.

a. $f(x) = e^x(11x^3 - 4x^{-1})$ for $x \neq 0$

b. $f(x) = (7x^4 + 3x^{-2})(6x^3 - 4x)$ for $x \neq 0$

4. Find the equation of the tangent line to the curve which is the graph of the equation $y = x^{3/2}$ at the point $(4, 8)$.

b. Find the equation of the tangent line to the curve $y = x^{3/2}$ at the point where $x = 16$.

5. Find the equation of the tangent line to the curve which is the graph of the function $f(x) = x^2 + 4x^{1/2} + 6x^{3/2}$ at the point where $x = 4$.

6. A weather balloon is released and rises vertically such that its distance above the ground during the first 10 seconds of flight is given by $s(t) = 4 + 3t + t^2$ where $s(t)$ is in feet and t is in seconds. Find the velocity of the balloon at $t = 4$.

7. If $f(x) = x^{3/2}g(x)$, $g(4) = 5$, $g'(4) = 8$, $g(9) = 4$, and $g'(9) = -2$, find $f(4)$, $f'(4)$, $f(9)$ and $f'(9)$. Ans. $f'(4) = 79$.

8. If $f(x) = (x^3 + 5x)g(x)$, $g(1) = 5$, $g'(1) = 4$, $g(2) = -1$, and $g'(2) = 3$, find $f(1)$, $f'(1)$, $f(2)$ and $f'(2)$. Find the equation of the tangent line to the curve $y = f(x)$ at the point where $x = 2$.

6323 The Quotient Rule

Theorem. The Quotient Rule for Differentiation. Given two functions $f(x)$ and $g(x)$ for which $f'(x)$ and $g'(x)$ exist, then

$$\frac{d}{dx}\left[\frac{f(x)}{g(x)}\right] = \frac{g(x)\frac{d}{dx}[f(x)] - f(x)\frac{d}{dx}[g(x)]}{[g(x)]^2}.$$

The Quotient Rule in words: The derivative of the quotient of two functions is given by the denominator times the derivative of the numerator minus the numerator times the derivative of the denominator all over the denominator squared. The proof of this theorem is not difficult. The critical step in the proof of this quotient rule for differentiation makes use of the quotient rule for limits. When we studied limits earlier we did not study the quotient rule for limits by that name. When we were evaluating limits we knew that the quotient rule was true and how to make use of it. But without a formal statement of the quotient rule for limits we are unable to understand a detailed proof of the quotient rule for differentiation.

Example 1. Let $f(x) = \dfrac{x^3 + 5x}{7x^2 + 9}$, find the derivative $f'(x)$.

Solution. First, note that this is a quotient. This means that we begin by using the quotient rule. Replace $f(x)$ with $x^3 + 5x$ and $g(x)$ with $7x^2 + 9$, in the quotient rule, then the quotient rule says

$$\frac{d}{dx}\left[\frac{x^3 + 5x}{7x^2 + 9}\right] = \frac{(7x^2 + 9)\frac{d}{dx}(x^3 + 5x) - (x^3 + 5x)\frac{d}{dx}[7x^2 + 9]}{(7x^2 + 9)^2}.$$

Applying Theorems 1, 2, and 3 from the previous section where we learned to differentiate polynomials, we find the derivatives. The result is

$$\frac{(7x^2 + 9)(3x^2 + 5) - (x^3 + 5x)(14x)}{(7x^2 + 9)^2} =$$

$$\frac{21x^4 + 35x^2 + 27x^2 + 45 - 14x^4 - 70x^2}{(7x^2 + 9)^2}.$$

Simplifying, we get

$$f'(x) = \frac{7x^4 - 8x^2 + 45}{(7x^2 + 9)^2}.$$

47

Example 2. Find $\dfrac{d}{dx}\left[\dfrac{x^2 e^x}{5x+8}\right]$.

Solution. Replacing $f(x)$ with $x^2 e^x$ and $g(x)$ with $5x+8$ in the quotient rule, we get

$$\frac{d}{dx}\left[\frac{x^2 e^x}{5x+8}\right] = \frac{(5x+8)\frac{d}{dx}(x^2 e^x) - (x^2 e^x)\frac{d}{dx}(5x+8)}{(5x+8)^2}$$

In order to find $\dfrac{d}{dx}(x^2 e^x)$ we use the product rule which says

$$\frac{d}{dx}(x^2 e^x) = x^2 \frac{d}{dx}(e^x) + e^x \frac{d}{dx}(x^2) = x^2 e^x + e^x(2x).$$

Making use of this result, we have

$$\frac{d}{dx}\left[\frac{x^2 e^x}{5x+8}\right] = \frac{(5x+8)(x^2 e^x + 2xe^x) - (x^2 e^x)(5)}{(5x+8)^2}$$

$$= \frac{5x^3 e^x + 8x^2 e^x + 10x^2 e^x + 16xe^x - 5x^2 e^x}{(5x+8)^2}$$

$$= \frac{5x^3 e^x + 13x^2 e^x + 16xe^x}{(5x+8)^2}.$$

Example 3. Let $f(x) = \dfrac{g(x)}{3x^2+8}$. Suppose $g(3) = 5$ and $g'(3) = 4$, find $f'(3)$.

Solution. Note that this is a quotient and so we start by using the quotient rule.

$$f'(x) = \frac{d}{dx}\left[\frac{g(x)}{3x^2+8}\right] = \frac{(3x^2+8)\frac{d}{dx}[g(x)] - g(x)\frac{d}{dx}[3x^2+8]}{(3x^2+8)^2}$$

$$= \frac{(3x^2+8)g'(x) - (6x)g(x)}{(3x^2+8)^2}.$$

Replacing x with 3, we get

$$f'(3) = \frac{(3\cdot 9+8)g'(3) - (6\cdot 3)g(3)}{(3\cdot 9+8)^2}$$

$$= \frac{35g'(3) - 18g(3)}{1225}.$$

Substituting in the values $g(3) = 5$ and $g'(3) = 4$ we get

$$f'(3) = \frac{35(4) - 18(5)}{1225} = \frac{50}{1225} = \frac{2}{49}.$$

Example 4. Let $f(x) = \dfrac{x^2 g(x)}{5x^2 + 7}$. Suppose that $g(2) = 6$ and $g'(2) = -2$, find $f'(2)$.

Solution. This fraction type function indicates that $f(x) = [x^2 g(x)] \div [5x^2 + 7]$, that is, the function $f(x)$ is the quotient of two other functions. Therefore, we must apply the quotient rule for differentiating the quotient of two functions. Applying the quotient rule gives

$$f'(x) = \frac{d}{dx}\left[\frac{x^2 g(x)}{5x^2 + 7}\right] = \frac{(5x^2 + 7)\frac{d}{dx}[x^2 g(x)] - [x^2 g(x)]\frac{d}{dx}(5x^2 + 7)}{(5x^2 + 7)^2}.$$

Applying the product rule for differentiating to the product $x^2 g(x)$, we find the derivative:

$$\frac{d}{dx}[x^2 g(x)] = x^2 \frac{d}{dx}[g(x)] + g(x)\frac{d}{dx}(x^2) = x^2 g'(x) + 2x g(x).$$

Substituting this into the expression for $f'(x)$, we have

$$f'(x) = \frac{(5x^2 + 7)[x^2 g'(x) + 2x g(x)] - x^2 g(x)[10x]}{(5x^2 + 7)^2}$$

$$= \frac{x^2(5x^2 + 7)g'(x) + (14x)g(x)}{(5x^2 + 7)^2}.$$

Replacing x with 2, we get

$$f'(2) = \frac{4(5 \cdot 4 + 7)g'(2) + 14(2)g(2)}{(5 \cdot 4 + 7)^2}$$

$$= \frac{108 g'(2) + 28 g(2)}{729}.$$

Substituting the values $g'(2) = -2$ and $g(2) = 6$, we get

$$f'(2) = \frac{108(-2) + 28(6)}{729} = -\frac{48}{729} = -\frac{16}{243}.$$

49

Exercises

1. Use the quotient rule to find the derivative of the following functions:

a. $f(x) = \dfrac{3x^2 + 7x}{x^3 + 10}$

   b. $g(x) = \dfrac{8x^2 + 5}{x^2 e^x}$

2. Let $f(x) = \dfrac{5x^2 + 2}{g(x)}$. Suppose that $g(2) = 3$ and $g'(2) = 5$, find $f'(2)$.

3. Suppose $f(x) = \dfrac{g(x)}{5x^2 - 4x}$ for $x \neq 0$ and $x \neq 4/5$. If $g(2) = 5, g(6) = -3, g(12) = 8$, $g'(2) = 4$, $g'(6) = 7$ and $g'(12) = 15$, find $f(2)$ and $f'(2)$. Ans. $f(2) = 5/12$ and $f'(2) = -2/9$.

4. Let $f(x) = \dfrac{(3x + 5)g(x)}{4x^2 + 9}$. Suppose that $g(3) = 5/9$ and $g'(3) = 2/7$, find $f'(3)$.

5. Let $f(x) = \dfrac{x}{g(x)}$. Suppose that $g(3) = 6$, $g(6) = 4$, $g'(3) = 5$, and $g'(6) = -3$, find $f'(3)$.

6325 Derivatives of Trigonometric Functions

There are six standard trigonometric functions. They are $\sin x$, $\cos x$, $\tan x$, $\cot x$, $\sec x$, and $\csc x$. We would like to find the derivatives of these functions. Let us start by finding the derivative of the function $\sin x$. We have no choice but to try to find the derivative of $\sin x$ using the definition of derivative. In order to find the derivative of $\sin x$ we need the value of a couple of limits.

Lemma 1. $\displaystyle\lim_{h\to 0} \frac{\sin(h)}{h} = 1.$

We prove that this limit is 1 using a geometric argument. The proof that the value of this limit is one is a rather long geometric argument but only requires a good knowledge of trigonometry. We will not give the proof here. In order for $\displaystyle\lim_{h\to 0} \frac{\sin(h)}{h} = 1$ to be true the number h must be measured in radians. This limit is of fundamental importance. Let $f(x) = \sin x$, then Lemma 1 enables us to find $f'(0)$. The definition of derivative says that

$$f'(0) = \lim_{h\to 0} \frac{\sin(h) - \sin(0)}{h - 0} = \lim_{h\to 0} \frac{\sin(h)}{h} = 1.$$

Thus, if $f(x) = \sin x$, then $f'(0) = 1$. If $f(x) = \sin x$, we still need to find $f'(x)$ for other values of x.

Lemma 2. $\displaystyle\lim_{h\to 0} \frac{\cos(h) - 1}{h} = 0.$

Proof. Note that substituting $h = 0$ gives $0/0$. This is an indeterminate form. In order to evaluate this limit we rewrite the fraction as follows:

$$\frac{\cos(h) - 1}{h} = \frac{[\cos(h) - 1][\cos(h) + 1]}{h[\cos(h) + 1]} = \frac{\cos^2(h) - 1}{h[\cos(h) + 1]}$$

$$= \frac{-\sin^2(h)}{h[\cos(h) + 1]} = \frac{\sin(h)}{h} \cdot \frac{-\sin(h)}{\cos(h) + 1}.$$

The function $\dfrac{\sin(h)}{\cos(h) + 1}$ is continuous at $h = 0$ and so the limit as h approaches zero is $\dfrac{\sin(0)}{\cos(0)+1} = \dfrac{0}{2} = 0$. It follows that

$$\lim_{h \to 0} \frac{\cos(h) - 1}{h} = \left[\lim_{h \to 0} \frac{\sin(h)}{h}\right] \left[\lim_{h \to 0} \frac{-\sin(h)}{\cos(h) + 1}\right]$$
$$= 1 \cdot 0 = 0.$$

Theorem 1. If $f(x) = \sin x$, then $f'(x) = \cos x$.

Proof. Let $f(x) = \sin x$, then the definition of derivative says that the derivative of $f(x)$ at $x = b$ is

$$f'(b) = \lim_{h \to 0} \frac{\sin(b + h) - \sin(b)}{h}.$$

Recall the trigonometric identity $\sin(b + h) = \sin(b)\cos(h) + \sin(h)\cos(b)$. This identity enables us to write the derivative of $\sin x$ at $x = b$ as

$$f'(b) = \lim_{h \to 0} \frac{\sin(b)\cos(h) + \sin(h)\cos(b) - \sin(b)}{h}$$
$$= \lim_{h \to 0} \frac{\sin(b)[\cos(h) - 1] + \cos(b)[\sin(h)]}{h}$$
$$= \lim_{h \to 0} \frac{\sin(b)[\cos(h) - 1]}{h} + \lim_{h \to 0} \frac{\cos(b)\sin(h)}{h}$$
$$= \sin(b) \lim_{h \to 0} \frac{[\cos(h) - 1]}{h} + \cos(b) \lim_{h \to 0} \frac{\sin(h)}{h}.$$

We evaluate these limits using Lemma 2 and Lemma 1. The result is

$$f'(b) = [\sin(b)][0] + [\cos(b)][1] = \cos b.$$

We would write this result as $f'(x) = \cos x$ or

$$\frac{d}{dx}[\sin(x)] = \cos(x).$$

Theorem 2. Let $g(x) = \cos x$, then $g'(x) = -\sin x$.

Proof. Let $g(x) = \cos x$, then the definition of derivative says that the derivative of $g(x)$ at $x = b$ is given by

$$g'(b) = \lim_{h \to 0} \frac{\cos(b + h) - \cos(b)}{h}.$$

Recall the trigonometric identity

$$\cos(b + h) = \cos(b)\cos(h) - \sin(b)\sin(h).$$

This identity enables us to write the derivative of $\cos x$ at $x = b$ as

$$g'(b) = \lim_{h \to 0} \frac{\cos(b)\cos(h) - \sin(b)\sin(h) - \cos(b)}{h}$$
$$= \lim_{h \to 0} \frac{\cos(b)[\cos(h) - 1] - \sin(b)[\sin(h)]}{h}$$
$$= [\cos(b)] \lim_{h \to 0} \frac{\cos(h) - 1}{h} - [\sin(b)] \lim_{h \to 0} \frac{\sin(h)}{h}$$

We use Lemma 2 and Lemma 1 to evaluate these limits. The result is

$$g'(b) = [\cos(b)](0) - [\sin(b)](1)$$
$$= -\sin(b).$$

We would usually write this result as

$$g'(x) = -\sin x \quad \text{or} \quad \frac{d}{dx}[\cos(x)] = -\sin(x).$$

In order to find the derivative of $\tan x$ we express $\tan x$ using the identity $\tan x = \dfrac{\sin x}{\cos x}$ and use the quotient rule.

$$\frac{d}{dx}[\tan x] = \frac{d}{dx}\left[\frac{\sin x}{\cos x}\right] = \frac{(\cos x)\frac{d}{dx}(\sin x) - (\sin x)\frac{d}{dx}(\cos x)}{\cos^2 x}$$
$$= \frac{\cos^2 x + \sin^2 x}{\cos^2 x} = \frac{1}{\cos^2 x} = \sec^2 x.$$

In order to find the derivative of $\sec x$ we express $\sec x$ using the identity $\sec x = \dfrac{1}{\cos x}$ and use the quotient rule.

$$\frac{d}{dx}[\sec x] = \frac{d}{dx}\left[\frac{1}{\cos x}\right] = \frac{(\cos x)(0) - (1)\frac{d}{dx}(\cos x)}{\cos^2 x}$$

$$= \frac{\sin x}{\cos^2 x} = \frac{\sin x}{\cos x}\frac{1}{\cos x} = \tan x \sec x.$$

Using similar methods

$$\frac{d}{dx}[\cot x] = -\csc^2 x \quad \text{and} \quad \frac{d}{dx}[\csc x] = -(\cot x)(\csc x)$$

We have the following table of derivatives

$$\frac{d}{dx}[\sin x] = \cos x$$

$$\frac{d}{dx}[\cos x] = -\sin x$$

$$\frac{d}{dx}[\tan x] = \sec^2 x$$

$$\frac{d}{dx}[\cot x] = -\csc^2 x$$

$$\frac{d}{dx}[\sec x] = (\sec x)(\tan x)$$

$$\frac{d}{dx}[\csc x] = -(\csc x)(\cot x).$$

Example 1. Let $y = x\cos x - 8\sec x + 5\cot x$, find $\frac{dy}{dx}$.

Solution. Using the product rule to differentiate $x\cos x$, we have

$$\frac{dy}{dx} = \frac{d}{dx}[x\cos x - 8\sec x + 5\cot x]$$

$$= x\frac{d}{dx}[\cos x] + \cos x\frac{d}{dx}[x] - 8\frac{d}{dx}[\sec x] + 5\frac{d}{dx}[\cot x]$$

$$= -x\sin x + \cos x - 8\tan x \sec x - 5\csc^2 x.$$

Example 2. Let $g(x) = (\tan x)(1 + \sin x)$, find $g'(x)$.

Solution. Since this function is expressed as a product, we begin by using the product rule.

$$g'(x) = (\tan x)\frac{d}{dx}(1 + \sin x) + (1 + \sin x)\frac{d}{dx}(\tan x)$$
$$= (\tan x)(\cos x) + (1 + \sin x)(\sec^2 x)$$
$$= \sin x + \sec^2 x + \tan x \sec x.$$

Example 3. Let $f(x) = \dfrac{x + \sin x}{x^2 + \cos x}$, find $f'(x)$.

Solution. Since this function is expressed as a quotient the first step must be the quotient rule.

$$f'(x) = \frac{(x^2 + \cos x)\frac{d}{dx}(x + \sin x) - (x + \sin x)\frac{d}{dx}(x^2 + \cos x)}{(x^2 + \cos x)^2}$$
$$= \frac{(x^2 + \cos x)(1 + \cos x) - (x + \sin x)(2x - \sin x)}{(x^2 + \cos x)^2}$$
$$= \frac{-x^2 + \cos x + x^2 \cos x - x \sin x + 1}{(x^2 + \cos x)^2}.$$

Exercises

Find the derivative of each of the following functions:

1. $f(x) = x \sin x + 8 \cos x - 11 \tan x$

2. $g(x) = 3 \sec x - 2\csc x + 12 \cot x$

3. $f(x) = (1 + \cos x)(x + \sin x)$

4. $g(x) = \sec x(1 + \tan x)$

5. $f(x) = \dfrac{\tan x + \sin x}{x + \cos x}$.

6. Find the equation of the tangent line to the curve $y = \sec x + 2 \cos x$ at the point $(\pi/3, 3)$. $y = 3 + \sqrt{3}[x - (\pi/3)]$.

55

6327 Composite Functions

So far we have found the derivatives for some elementary functions. Some examples of elementary functions are $x^5, x^{1/2}, e^x$ and $\sin x$. Not all the functions we encounter in calculus are elementary functions. We can form more complicated functions by taking sums, products, and quotients of these elementary functions. Even more complicated functions can be constructed by placing one elementary function inside another elementary function. Such functions are called composite functions. Composite functions are very common, important, and somewhat hard to understand. Some examples of composite functions are: $(5x^2 + 11)^5$, $e^{\sin x}$, $\sqrt{5x^2 + 11}$, and $(\sin x)^2$. The function $\sqrt{5x^2 + 11}$ is constructed by inserting the function $5x^2 + 11$ in the place of x in the function $\sqrt{x}$. The composite function $e^{\sin x}$ is constructed by inserting the function $\sin x$ in place of x in the function e^x. In order to make the idea of composite functions clearer let us introduce some notation.

Example 1. Let $g(x) = x^3$ and $h(x) = \cos x$. If we replace x in $g(x) = x^3$ with $h(x) = \cos x$, then we obtain the composite function $g(h(x)) = (\cos x)^3$. On the other hand, if we reverse the order and replace x in $h(x) = \cos x$ with $g(x) = x^3$, then we obtain the composite function $h(g(x)) = \cos(x^3)$. The notation $h(g(x))$ indicates that we replace x in the function $h(x)$ with the function $g(x)$.

Example 2. Let $g(x) = \sin x$ and $h(x) = x^{1/3}$. If we replace x in $g(x) = \sin x$ with $h(x) = x^{1/3}$, we obtain the composite function $g(h(x)) = \sin(x^{1/3})$. On the other hand, if we replace x in $h(x) = x^{1/3}$ with $g(x) = \sin x$, we obtain the composite function $h(g(x)) = (\sin x)^{1/3}$.

Once we understand what composite functions are, then the important thing is to be able to decompose a composite function into its component functions.

Example 3. Decompose the composite function $f(x) = (\cos x)^4$ into its component functions.

Solution. The composite function $f(x) = (\cos x)^4$ is a composite of $g(x) = x^4$ and $h(x) = \cos x$, because we get $(\cos x)^4$ by replacing x in $g(x) = x^4$ with $\cos x$. The composite function is $g(h(x)) = (\cos x)^4$.

Example 4. Decompose the composite function e^{9x} into its component functions.

Solution. The component functions are $g(x) = e^x$ and $h(x) = 9x$ because we obtain e^{9x} by replacing x in $g(x) = e^x$ with $9x$, that is, $g(h(x)) = e^{9x}$.

Example 5. Decompose the composite function $\sin(e^x)$ into its component functions.

Solution. The component functions are $g(x) = \sin x$ and $h(x) = e^x$ because we obtain $\sin(e^x)$ by replacing x in $g(x) = \sin x$ with $h(x) = e^x$, that is, $g(h(x)) = \sin(e^x)$.

Example 6. Composite functions are often made up of more than two component functions. Decompose the composite function $e^{(\cos x)^3}$ into its component functions.

Solution. Composite functions are broken down into component functions and these component functions must always be elementary functions. The component functions are $g(x) = e^x$, $h(x) = x^3$ and $m(x) = \cos x$. We replace x in $h(x) = x^3$ with $\cos x$ which gives $h(m(x)) = (\cos x)^3$. We then replace x in $g(x) = e^x$ with $h(m(x)) = (\cos x)^3$ to obtain $g(h(m(x))) = e^{(\cos x)^3}$.

Example 7. Decompose the composite function $e^{(\sin x^3)^2}$ into its component functions.

Solution. In order to determine the component functions of $f(x) = e^{(\sin x^3)^2}$ it can be helpful to consider the steps we would use to calculate $f(2)$. The first step is to calculate 2^3 and so the intermost function is $p(x) = x^3$. The next step is to calculate $\sin(2^3) = \sin(8)$ and so the next component function is $m(x) = \sin x$. Next calculate $(\sin 8)^2$ and so the next component function is $h(x) = x^2$. Finally, calculate $e^{(\sin 8)^2}$ and so the outer most component function is $g(x) = e^x$. Thus, the component functions are $g(x) = e^x$, $h(x) = x^2$, $m(x) = \sin x$, and $p(x) = x^3$. Note that $g(x)$, $h(x)$, $m(x)$ and $p(x)$ are all elementary functions. We replace x in $m(x) = \sin x$ with $p(x) = x^3$ to get $m(p(x)) = \sin x^3$. We replace x in $h(x) = x^2$ with $m(p(x)) = \sin x^3$ to get $h(m(p(x))) = (\sin x^3)^2$. Finally, we replace x in

$g(x) = e^x$ with $h(m(p(x))) = (\sin x^3)^2$ to obtain $g(h(m(p(x)))) = e^{(\sin x^3)^2}$. This is a composite function with four layers.

Example 8. Decompose the following composite function into its component functions:

$$f(x) = [x^3 + \sqrt{4x^2 + 9}]^{2/3}.$$

Solution. Start with the outside function. Let $g(x) = x^{2/3}$. There is something different about this example. The first inside function is $x^3 + \sqrt{4x^2 + 9}$. This is a sum. Now x^3 is an elementary function but $\sqrt{4x^2 + 9}$ is a composite function. Let $h(x) = x^{1/2}$ and $m(x) = 4x^2 + 9$, then $\sqrt{4x^2 + 9}$ is the composite function $h(m(x)) = \sqrt{4x^2 + 9}$. Using this notation, we have

$$f(x) = g(x^3 + h(m(x))) = [x^3 + \sqrt{4x^2 + 9}]^{2/3}.$$

Exercises

Decompose the following composite functions into their component functions.

1. $e^{\sin x}$

2. $\tan 8x$

3. $\sqrt{8x^2 + 14}$

4. $\tan(3x^2)$

5. $\sin(e^{6x})$

6. $[\cos(7x^2)]^{1/2}$, $0 \le x \le 0.2$

7. $e^{(\sin 8x)^2}$

8. $[10x + \cos^3 x]^{3/2}$

9. $[x^3 + \sqrt{5x^2 + 8}]^{1/3}$.

10. $[x^{1/2} + \cos(x)]^{4/3}$

6329 Chain Rule

In order to be able to differentiate composite functions we need the chain rule for differentiation.

Theorem. Chain Rule. Suppose we have the composite function $f(x) = g(h(x))$, then the derivative of this function is given by

$$f'(x) = g'(h(x))h'(x).$$

The rule is easier to understand if we denote the inside function $h(x)$ by a letter. Let $u = h(x)$, then the rule says

$$f'(x) = \frac{d}{du}[g(u)]\frac{du}{dx}.$$

It is important to be clear about what $g'(h(u))$ or $\frac{d}{du}(g(u))$ means. In practice the function $g(x)$ is an elementary function. The easy way to find $g'(h(x))$ is to say $u = h(x)$ is the independent variable. Replace $h(x)$ with u, find $g'(u)$, and then put back $h(x)$ for u. The chain rule is correct even if the functions $g(x)$ and $h(x)$ are not elementary functions.

Example 1. Use the chain rule to find the derivative of the composite function $f(x) = e^{\sin x}$.

Solution. The inside function is $u = h(x) = \sin x$. The outside function is e^x. We can express $f(x)$ as $f(x) = e^u$ where $u = \sin x$. The chain rule says:

$$f'(x) = \frac{d}{du}(e^u)\frac{du}{dx} = e^u\frac{d}{dx}(\sin x) = e^u\cos x.$$

Replacing u with $\sin x$, we get

$$f'(x) = e^{\sin x}\cos x.$$

Example 2. Let $g(x) = \sin 8x$, find $g'(x)$.

Solution. This is a composite function. The inner function is $8x$. Let $u = 8x$. The outer function is $\sin u$. The chain rule says that

$$g'(x) = \frac{d}{du}(\sin u)\frac{du}{dx} = (\cos u)\frac{d}{dx}(8x) = 8\cos u.$$

We should replace u with $8x$ to get

$$g'(x) = 8 \cos 8x.$$

Example 3. Let $f(x) = (2 + \cos x)^{1/2}$, find $f'(x)$.

Solution. This is a composite function. First, determine the outside function. The outside function must be an elementary function. Let u equal everything inside the outside most function. We do this no matter if this makes u an elementary function or not. Let $u = 2 + \cos x$, then the outside function is $u^{1/2}$. We can write $f(x) = u^{1/2}$. The chain rule says that

$$f'(x) = \frac{d}{du}(u^{1/2})\frac{du}{dx} = \frac{1}{2}u^{-1/2}\frac{d}{dx}(2 + \cos x)$$
$$= \frac{1}{2\sqrt{u}}(-\sin x) = \frac{-\sin x}{2\sqrt{2 + \cos x}}.$$

Example 4. Let $g(x) = [x^2 + \sin(x^3)]^{3/2}$, find $g'(x)$.

Solution. This is a composite function. It is a three layer function. The inner most function is x^3, the middle function is a sum, and the outer function is $x^{3/2}$. However, when we have a multilayer function like this one it is best to begin without being concerned about what all the component functions are. With a multilayer function it is best to just consider one layer at a time. We start by identifying the outer most function. We start by letting u equal the function which is inside the outermost function. The outer function is $[\ ]^{3/2}$. We can write the given function $g(x)$ as $g(x) = u^{3/2}$ where $u = x^2 + \sin(x^3)$. The first inside function need not be an elementary function. We always use the substitution that u equals the first inside function. The chain rule says

$$g'(x) = \frac{d}{du}(u^{3/2})\frac{du}{dx} = \frac{3}{2}u^{1/2}\frac{du}{dx}.$$

Next we must turn our attention to the problem of finding the expression $\frac{du}{dx}$. The problem is that this time $u(x)$ is not an elementary function. We see that u involves a composite function, namely $\sin(x^3)$. The function x^3 is an elementary function, but in order to find the derivative of $\sin(x^3)$ we

60

must treat $\sin(x^3)$ as a composite function. In order to find $\frac{d}{dx}[x^2 + \sin(x^3)]$ we must find the derivative of the composite function $\sin(x^3)$. We find the derivative the same way we find the derivative of any other composite function. Let $w = x^3$, then $\sin(w) = \sin(x^3)$. The chain rule says

$$\frac{d}{dx}[\sin(x^3)] = \frac{d}{dw}[\sin w]\frac{dw}{dx} = \cos(w)\frac{d}{dx}(x^3) = \cos(x^3)(3x^2).$$

Using this fact, we have

$$\frac{du}{dx} = \frac{d}{dx}[x^2 + \sin(x^3)] = 2x + 3x^2\cos(x^3).$$

Substituting back

$$g'(x) = \frac{3}{2}u^{1/2}\frac{du}{dx} = \frac{3}{2}[x^2 + \sin(x^3)]^{1/2}[2x + (3x^2)\cos(x^3)].$$

Example 5. Let $f(x) = [\sin(x^2 + e^{10x})]^4$, find $f'(x)$.

Solution. This is a multilayered composite function. We start by looking at the outer function only. We let u equal to the function inside the first outer function. We do this even when the result is that u is also a composite function. Let $u = \sin(x^2 + e^{10x})$, then we can express this function as $f(x) = u^4$. Applying the chain rule, we get

$$f'(x) = \frac{df}{du}\frac{du}{dx} = \frac{d}{du}(u^4)\frac{du}{dx} = 4u^3\frac{du}{dx}.$$

We need to find $\frac{du}{dx}$. The function $u = \sin(x^2 + e^{10x})$ is a many layered composite function. The outer function is $\sin(\ )$ and the inside function is $x^2 + e^{10x}$. Let $w = x^2 + e^{10x}$, then $u = \sin w$. Applying the chain rule

$$\frac{du}{dx} = \frac{du}{dw}\frac{dw}{dx} = \frac{d}{dw}(\sin w)\frac{dw}{dx} = (\cos w)\frac{dw}{dx}.$$

Will it never end. The function w involves a composite function since $w = x^2 + e^{10x}$. The function x^2 is an elementary function, but e^{10x} is a composite function. In order to find $\frac{d}{dx}[e^{10x}]$ let $y = 10x$, then the chain rule says

$$\frac{d}{dx}(e^y) = \frac{d}{dy}(e^y)\frac{dy}{dx} = e^y\frac{d}{dx}(10x) = 10e^y = 10e^{10x}.$$

Using this, we get

$$\frac{dw}{dx} = \frac{d}{dx}(x^2 + e^y) = \frac{d}{dx}(x^2) + \frac{d}{dx}(e^y) = 2x + 10e^{10x}.$$

It follows that

$$\frac{du}{dx} = (\cos w)(2x + 10e^{10x}) = [\cos(x^2 + e^{10x})](2x + 10e^{10x}).$$

Substituting back

$$f'(x) = 4u^3 \frac{du}{dx} = 4[\sin(x^2 + e^{10x})]^3 \frac{du}{dx}$$
$$= 4[\sin(x^2 + e^{10x})]^3 [\cos(x^2 + e^{10x})](2x + 10e^{10x}).$$

Example 6. Let $f(x) = [\sqrt{x^2 + 4} + \cos x]^{2/3}$. Use the chain rule to find $f'(x)$.

Solution. Clearly $f(x)$ is a composite function and so requires the chain rule. We start with the outermost function. Let u equal to the function inside the first outermost function. Let $f(x) = u^{2/3}$, where $u = \sqrt{x^2 + 4} + \cos x$. Using the chain rule

$$\frac{df}{dx} = \frac{df}{du}\frac{du}{dx} = \frac{d}{du}(u^{2/3})\frac{du}{dx} = \frac{2}{3}u^{-1/3}\frac{du}{dx}.$$

The function $u(x)$ is the sum of two functions. The function $\cos x$ is an elementary function, but the function $\sqrt{x^2 + 4}$ is a composite function.

$$\frac{du}{dx} = \frac{d}{dx}[\sqrt{x^2 + 4} + \cos x] = \frac{d}{dx}[\sqrt{x^3 + 4}] - \sin x.$$

We again need the chain rule to find $\frac{d}{dx}[\sqrt{x^2 + 4}]$. Let $w = \sqrt{x^2 + 4}$ and $z = x^2 + 4$, then $w = z^{1/2}$. Applying the chain rule, we get

$$\frac{dw}{dx} = \frac{dw}{dz}\frac{dz}{dx} = \frac{1}{2}z^{-1/2}\frac{d}{dx}(x^2 + 4) = z^{-1/2}(x) = \frac{x}{\sqrt{z}}.$$

Thus,

$$\frac{du}{dx} = \frac{x}{\sqrt{z}} - \sin x = \frac{x}{\sqrt{x^2 + 4}} - \sin x.$$

Therefore,

$$\frac{df}{dx} = \frac{2}{3} \left[\sqrt{x^2 + 4} + \cos x \right]^{-1/3} \left[\frac{x}{\sqrt{x^2 + 4}} - \sin x \right].$$

When faced with writing down the derivative of a composite function we often try to write down as little as possible. After deciding which function is inside the first outside function instead of writing $u(x)$ equal to this inside function, we try to keep the function $u(x)$ in our head. This works pretty well for simple composite functions, but usually leads to mistakes when working with a complicated composite function. For the composite function $\sin(3x^2)$ it is easy to do the chain rule using $u = 3x^2$ without writing $u(x) = 3x^2$. For the composite function $[\sin(x^3 + \sqrt{x^2 + 1})]^{3/2}$ to try to do the chain rule without writing down the expression for $u(x)$ usually leads to mistakes.

Example 7. Let $f(x) = g(3x^2)$. Suppose that $g(2) = 4$, $g(12) = -3$, $g'(2) = 8$, and $g'(12) = 5$, find $f'(2)$.

Solution. The function $f(x)$ is a composite function. The first outside function is g and the inner function is $u = 3x^2$. The chain rule is

$$\frac{df}{dx} = \frac{dg}{dx} = \frac{dg}{du} \frac{du}{dx} = g'(u) \frac{d}{dx}(3x^2) = 6x g'(u).$$

This says that

$$f'(x) = 6x g'(3x^2).$$

Replace x with 2 in this equation and we get

$$f'(2) = 12 g'(12).$$

Since $g'(12) = 5$, it follows that

$$f'(2) = 12(5) = 60.$$

Example 8. Let $f(x) = \dfrac{g(3x^2 - 4x)}{x^2 + 4}$. Suppose that $g(2) = -4$, $g(4) = 16$, $g(8) = 11$, $g'(2) = 3$, $g'(4) = 5$, and $g'(8) = 10$, find $f'(2)$.

Solution. Note that $f(x)$ is the quotient of two functions. We begin by using the quotient rule to find the derivative of $f(x)$:

$$f'(x) = \frac{(x^2 + 4)\frac{d}{dx}[g(3x^2 - 4x)] - g(3x^2 - 4x)(2x)}{(x^2 + 4)^2}.$$

In order to find $\frac{d}{dx}[g(3x^2 - 4x)]$, we must use the chain rule. The outside function is $g(x)$ and the function inside this function is $3x^2 - 4x$. Therefore, we let $u = 3x^2 - 4x$. The chain rule says

$$\frac{d}{dx}[g(u)] = \frac{d}{du}[g(u)]\frac{du}{dx} = g'(u)(6x - 4)$$
$$= g'(3x^2 - 4x)(6x - 4).$$

Substituting back into the expression for $f'(x)$, we get

$$f'(x) = \frac{(x^2 + 4)(6x - 4)g'(3x^2 - 4x) - 2xg(3x^2 - 4x)}{(x^2 + 4)^2}.$$

Replacing x with 2, we get

$$f'(2) = \frac{(8)(8)g'(4) - 2(2)g(4)}{(8)^2} = \frac{16g'(4) - g(4)}{16}.$$

Substituting in $g(4) = 16$ and $g'(4) = 5$, we get

$$f'(2) = \frac{16(5) - 16}{16} = 4.$$

Example 9. Let $f(x) = [g(5x - 7)]^3$. Suppose that $g(3) = 15$, $g(5) = 11$, $g(8) = 3$, $g'(3) = 6$, $g'(5) = 7$, and $g'(8) = 4$, find $f'(3)$.

Solution. This function is a composite function of several layers. It is not a product or a quotient. The outer function is u^3, that is, $f(x) = u^3$ where $u = g(5x - 7)$. The first inside function is $g(5x - 7)$. We apply the chain rule

$$f'(x) = \frac{df}{dx} = \frac{d}{du}(u^3)\frac{du}{dx} = 3u^2\frac{du}{dx}.$$

Next, we need to find $\frac{du}{dx}$. The function $g(5x - 7)$ is a composite function. Again we need to use the chain rule. The outside function is $g(x)$ and the function inside this function is $5x - 7$. Let $w = 5x - 7$, then the chain rule says

$$\frac{du}{dx} = \frac{du}{dw}\frac{dw}{dx} = \frac{d}{dw}[g(w)]\frac{dw}{dx} = g'(w)(5).$$

It follows that

$$f'(x) = 3u^2 g'(w)(5)$$
$$= 15[g(5x - 7)]^2 g'(5x - 7).$$

Replacing x with 3, we get

$$f'(3) = 15[g(8)]^2 g'(8).$$

Using $g(8) = 3$ and $g'(8) = 4$, we get

$$f'(3) = 15[3]^2(4) = 540.$$

Example 10. Suppose $f(x) = \sin[g(\sqrt{4 + 3x^2})]$, then find $f'(x)$.

Solution. This is a composite function. The outermost function is $\sin()$ and the function inside this is $g(\sqrt{4 + 3x^2})$. Let $u = g(\sqrt{4 + 3x^2})$. The chain rule says

$$\frac{df}{dx} = \frac{df}{du}\frac{du}{dx} = \frac{d}{du}(\sin u)\frac{du}{dx} = \cos u \frac{du}{dx} = \cos[g(\sqrt{4 + 3x^2})]\frac{du}{dx}.$$

The function $u(x)$ is also composite. The outside function is $g(\ )$ and the function inside this function is $\sqrt{4 + 3x^2}$. Let $w = \sqrt{4 + 3x^2}$, then $u(x) = g(w)$ and

$$\frac{du}{dx} = \frac{du}{dw}\frac{dw}{dx} = g'(u)\frac{dw}{dx} = g'(\sqrt{4 + 3x^2})\frac{dw}{dx}.$$

The function $w(x)$ is a composite function. The outside function is the square root function and the function inside the square root is $4 + 3x^2$. Let $y(x) = 4 + 3x^2$, then $w = y^{1/2}$. Applying the chain rule, we get

$$\frac{dw}{dx} = \frac{dw}{dy}\frac{dy}{dx} = \frac{d}{dy}(y^{1/2})\frac{dy}{dx} = \frac{1}{2}y^{-1/2}(6x) = \frac{3x}{\sqrt{4 + 3x^2}}.$$

Thus

$$f'(x) = \cos[g(\sqrt{4+3x^2})]g'(\sqrt{4+3x^2})\frac{3x}{\sqrt{4+3x^2}}.$$

Exercises

Find the derivative of each of the following functions.

1. $f(x) = (8x^3 + bx)^4$

2. $F(x) = \sec^3 x + \tan^2 x$

3. $g(x) = (5x^2 + 8)^3 \sin 4x$

4. $f(x) = (2 + \sin bx)^3$

5. $g(x) = \dfrac{1 + \cos 4x}{(bx^2 + 8)^3}$

6. $F(x) = [x^2 + \sin^2 5x]^{1/2}$

7. $f(x) = \sin\sqrt{x^3 + 8x}$ for $x \geq 0$

8. $F(x) = e^{(\sin x^3)^2}$

9. $f(x) = \sqrt{\sin(10x) + e^{8x}}$ for $x \geq 0$

10. $g(x) = [x^3 + (5x^2 + 6)^{1/2}]^{4/3}$

11. Let $f(x) = g(5x^2 - 4)$. Suppose that $g(2) = -3$, $g(16) = 9$, $g(20) = 6$, $g'(2) = 7$, $g'(16) = 4$, $g'(20) = 5$, find $f'(2)$.

12. Let $f(x) = (3x^2 + 4)g(5x^2)$. Suppose that $g(2) = 6$, $g(20) = 5$, $g'(2) = 10$ and $g'(20) = 8$, find $f'(2)$.

13. Let $f(x) = \dfrac{x^2 + 6}{g(x^2)}$. Suppose that $g(3) = -4$, $g(6) = 8$, $g(9) = 5$, $g'(3) = 3$, $g'(6) = 11$, and $g'(9) = 4$, find $f'(3)$.

14. Let $f(x) = [x^2 + g(8x)]^{3/2}$. Suppose that $g(2) = 4$, $g(8) = 7$, $g(16) = 21$, $g'(2) = 10$, $g'(8) = 9$, $g'(16) = 5$, find $f'(2)$. Suppose $x^2 + g(8x) > 0$.

15. Let $f(x) = \sin[g(3x^2 + 2)]$. Suppose that $g(2) = \pi$, $g(12) = \pi/4$, $g(14) = \pi/3$, $g'(2) = 4$, $g'(12) = 10$ and $g'(14) = 6$, find $f'(2)$.

6335 Implicit Functions

All the functions that we have studied up to this time have been given by an explicit formula. When a function is given by an explicit formula we sometimes call them explicit functions. Some examples of explicit functions are $f(x) = \sin x$ and $g(x) = x^2 + xe^x$. Given a value of x we are able to directly compute the corresponding value of the function. We are now going to look at a few functions defined in another way. We are going to look at functions defined by an equation and one or more side conditions. Functions defined in this way are usually called implicit functions. It would be clearer what type of function we are discussing if we used the name "functions defined using an implicit method" rather than just the simple term "implicit functions".

We start with a simple implicit function. Let us consider "the implicit function defined by the equation $y^2 - 8y + x - 9 = 0$". The statement in quotes is the same sort of statement that we usually see as the definition of an implicit function, but it is not a complete definition. First, what this statement really means is the following: "Let y be defined as a function of x by the equation $y^2 - 8y + x - 9 = 0$". This means that given a value of the independent variable x, we find the corresponding value of the dependent variable y by substituting the value of x into the equation and solving for the corresponding value of y. Recall that in order to have a function there must correspond only one value of y to a given value of x. Let $x = 9$. Substituting $x = 9$ into $y^2 - 8y + x - 9 = 0$ gives $y^2 - 8y = 0$. The solutions of this equation are $y = 0$ and $y = 8$. Corresponding to the value $x = 9$ there are two values of y, namely, $y = 0$ and $y = 8$. In order to have a function we must make a choice between these two values of y. In order to define a function we must say something in addition to saying "given a value of x the corresponding value of y is a solution of the equation $y^2 - 8y + x - 9 = 0$." We will not try to prove it at the moment but if we add the requirement that $y \geq 4$, then there is only one value of y corresponding to each value of x. In other words the following statement does define a function: "Let $y = f(x)$ denote the function such that given x the corresponding value of y is the solution of the equation $y^2 - 8y + x - 9 = 0$ for which $y \geq 4$". More work would show that this function is only defined for $x \leq 25$. A different function is defined by the statement: "Let $y = g(x)$ denote the function such that given x the corresponding value of y is the solution of the equation $y^2 - 8y + x - 9 = 0$ such that $y \leq 4$". This function $g(x)$ is also only defined for $x \leq 25$.

Since $y^2 - 8y + x - 9 = 0$ is a relatively simple equation we can actually solve the equation to obtain an explicit expression for the two functions. The equation can be written as

$$y^2 - 8y = -x + 9$$
$$(y - 4)^2 = 25 - x$$
$$y = 4 \pm \sqrt{25 - x}.$$

The two functions are $f(x) = 4 + \sqrt{25 - x}$ and $g(x) = 4 - \sqrt{25 - x}$. These formulas define the functions $f(x)$ and $g(x)$ as explicit functions. From these explicit expression for $f(x)$ and $g(x)$ it is clear that $f(x) \geq 4$, $g(x) \leq 4$ and both are defined for $x \leq 25$.

Example 1. Let us consider the implicit function defined by the equation $y^2 + 4x^2 - 6y - 40x + 9 = 0$. What is the value of y corresponding to $x = 1$? Substituting $x = 1$ into the equation, we get

$$y^2 - 6y - 27 = 0$$
$$(y - 9)(y + 3) = 0.$$

The solutions are $y = 9$ and $y = -3$. Since there are two values of y the equation $y^2 + 4x^2 - 6y - 40x + 9 = 0$ must define two functions. The best way to understand the situation is to look at the graph of the equation $y^2 + 4x^2 - 6y - 40x + 9 = 0$. The graph of this equation is an ellipse and is given below. In order to graph this ellipse we would usually complete the square as follows:

$$4x^2 - 40x + y^2 - 6y = -9$$
$$4(x^2 - 10x + \quad) + (y^2 - 6y + \quad) = -9$$
$$4(x^2 - 10x + 25) + (y^2 - 6y + 9) = -9 + 100 + 9$$
$$4(x - 5)^2 + (y - 3)^2 = 100$$

The center of the ellipse is $(5, 3)$. From this equation we see that points on the ellipse are $(0, 3)$, $(10, 3)$, $(5, -7)$ and $(5, 13)$.

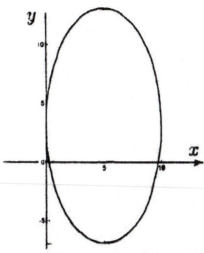

The first function is: Given a value of x the corresponding value of $y = f(x)$ is the solution of the equation $y^2 + 4x^2 - 6y - 40x + 9 = 0$ such that $y \geq 3$. This function is only defined for $0 \leq x \leq 10$. The graph of this function is the top half of the ellipse. The second function is defined by the statement: Let $y = g(x)$ denote the function such that for a given x the corresponding value of y is the solution of the equation $y^2 + 4x^2 - 6y - 40x + 9 = 0$ such that $y \leq 3$. This function is also only defined for $0 \leq x \leq 10$. The graph of this function is the lower half of the ellipse.

It should be pointed out that when discussing implicit functions we do not go crazy. We will always assume that any implicit function we are discussing is continuous at all points where it is defined. What this comes down to is the following. Go back to the equation $y^2 + 4x^2 - 6y - 40x + 9 = 0$. From Example 1, we considered two functions defined implicitly by this equation. These two functions can be written as

$$f(x) = 3 + \sqrt{100 - 4(x - 5)^2} \text{ and } g(x) = 3 - \sqrt{100 - 4(x - 5)^2}.$$

These are the only two continuous functions defined by this equation for $-5 \leq x \leq 5$. Consider the non continuous function

$$h(x) = \begin{cases} 3 - \sqrt{100 - 4(x - 5)^2} \text{ for } 0 \leq x < 5 \\ 3 + \sqrt{100 - 4(x - 5)^2} \text{ for } 5 \leq x \leq 10. \end{cases}$$

69

The function $h(x)$ satisfies the equation $4(x-5)^2 + (y-3)^2 = 100$ for all x such that $0 \le x \le 10$. However, it is not continuous at $x = 5$. We could construct a function which is not continuous at several values of x. It is clear that to include consideration of such functions in our discussion is a waste of time. When, we say we have a function defined implicitly by an equation we will always mean only the functions that are continuous for all values of the independent variable.

Example 2. Let us consider the function defined by the equation $x = \sin y$ where x is the independent variable and y is the dependent variable. Let $x = \sqrt{2}/2$, what are the corresponding values of y. Some solutions of the equation $\sin y = \sqrt{2}/2$ are $y = \pi/4$, $y = -\pi/4$, $y = 7\pi/4$, $y = -7\pi/4$, and $y = 9\pi/4$. In order to have a function we need a unique value of y. Clearly the equation $x = \sin y$ alone is not enough to define an implicit function. In addition, we need some side condition that requires us to choose a single value of y corresponding to each value of x. Different rules for choosing the value of y give different functions. It is clear that we can define more than two continuous functions using the equation $x = \sin y$. If we look at the graph of the equation $x = \sin y$ we can easily select an additional condition which will yield a function. Let us list three possible functions.

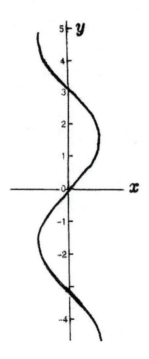

1. Let $y = f(x)$ denote the function such that given a value of x the corresponding value of y is the solution of the equation $x = \sin y$ such that $-\pi/2 \leq y \leq \pi/2$. This function is defined for $-1 \leq x \leq 1$. This function is usually given the name $y = \arcsin x$.

2. Let $y = g(x)$ denote the function such that given a value of x the corresponding value of y is the solution of the equation $x = \sin y$ such that $\pi/2 \leq y \leq 3\pi/2$. This function is defined for $-1 \leq x \leq 1$. Note that $g(x) = \pi - f(x)$.

3. Let $y = h(x)$ denote the function such that given a value of x the corresponding value of y is the solution of the equation $x = \sin y$ such that $-3\pi/2 < y < -\pi/2$. This function is defined for $-1 \leq x \leq 1$. Note that $h(x) = -\pi - f(x)$.

Previously we discussed the function e^x. The assumption was that we knew how to compute the values of this function for any x. However, a clear algorithm for computing e^x was not really defined. The definition of the number e were a little complicated. Since e is not a fraction (rational number) there is no method for finding e^x that involves only the usual operations of arithmetic. After we have discussed power series (this comes later) we could define e^x as

$$e^x = \sum_{n=0}^{\infty} \frac{x^n}{n!}.$$

Using facts about power series we could prove that e^y is continuous and that given a value of x with $x > 0$, the equation $x = e^y$ has exactly one solution.

Example 3. Let us look at another important implicit function. For a given value of y we know the meaning of e^y. Also recall that $e^y > 0$ for all y. In order to be able to solve the equation $x = e^y$, we must choose a value of x such that $x > 0$. The equation $x = e^y$ is the same as $x^{1/y} = e$. In order to get a positive number, namely e, using the power $x^{1/y}$, we must start with a number x such that $x > 0$. Given a number x, $x > 0$, there is only one number y such that $e^y = x$. Since there is only one value of y we do not need a side condition to tell us which value to choose.

Definition. Let $y = \log_e x = \ln x$ denote the function such that given a value of x with $x > 0$ the corresponding value of y is the solution of the equation $x = e^y$.

When we assume that $x > 0$ there is only one solution to the equation $x = e^y$ and so we do not need a side condition in order to get a function. We call this solution $y = \ln x$ the inverse function of e^x. When we use the notation $f(x) = e^x$ we sometimes use the rather poor notation $f^{-1}(x)$ for the inverse function $\ln(x)$. A function and its inverse have a unique relationship to each other. For the functions e^x and $\ln(x)$ this relationship is as follows. It is important to realize that $e^{(\ln x)} = x$ for all $x > 0$. This is obtained by replacing y in $x = e^y$ with $y = \ln x$. Also $\ln(e^x) = x$ for all real numbers x.

Theorem 1. $\ln(ab) = \ln(a) + \ln(b)$.

Proof. Let $N = \ln a$ and $P = \ln b$. These equations are the same as $a = e^N$ and $b = e^P$. It follows that $ab = (e^N)(e^P) = e^{N+P}$. This last equation is the same as $\ln(ab) = N + P$. It follows that $\ln(ab) = \ln(a) + \ln(b)$.

Theorem 2. $\ln a^x = x(\ln a)$.

Proof. Let $N = \ln a$, then $a = e^N$. It follows that $a^x = (e^N)^x = e^{xN}$. This equation is the same as $xN = \ln(a^x)$. Thus, $\ln(a^x) = xN = x(\ln a)$.

Theorem 3. $\ln(a/b) = \ln(a) - \ln(b)$.

Proof. $\ln(a/b) = \ln[a(b^{-1})] = \ln a + \ln(b^{-1}) = \ln a - \ln b$.

Given $a > 0$ let $y = a^x$, then $\ln y = \ln(a^x) = x(\ln a)$. This last equation is the same as $y = e^{x(\ln a)}$. Therefore,

Theorem 4. $a^x = e^{(\ln a)x}$.

This says that if we have the function a^x for any positive base a, we can express this function as $e^{(\ln a)x}$ using the base e. In particular, $10^x = e^{(\ln 10)x}$. When given the function 10^x we will convert it to the notation $e^{(\ln 10)x}$.

We can also define the following functions

$$a^{[h(x)]} = e^{(\ln a)[h(x)]} \text{ for } a > 0.$$

$$[g(x)]^{[h(x)]} = e^{[\ln g(x)][h(x)]} \text{ for } g(x) > 0.$$

Since we now know the chain rule we can find the derivative of such functions.

Exercises

1. Two continuous functions are defined implicitly by the equation $y^2 - 6y - 2x + 1 = 0$. If $x = -2$, what are the corresponding values of y? Find explicit expressions for these functions, that is, solve this equation for y. In order to solve you will need to complete the square. For what values of x are these functions defined? What is the range of values of y for these functions?

2. Suppose that for a given value of x the corresponding value of y is the solution of the equation $y^2 - 10y + 2x + 5 = 0$. If $x = 2$, what are the corresponding values of y? If $x = -8$, what are the corresponding values of y? Below is a graph of the equation $y^2 - 10y + 2x + 5 = 0$.

This is a parabola. We can define a continuous function as follows. Let $y = f(x)$ denote the function such that given a value of x the corresponding value of y is the solution of $y^2 - 10y + 2x + 5 = 0$ such that $y \geq 5$. For what values of x is the function defined? What is the graph of this function? State the definition of another continuous function defined by the equation $y^2 - 10y + 2x + 5 = 0$.

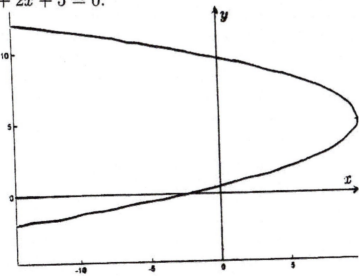

3. Consider the equation $x^2 + y^2 - 6x - 4y = 12$. What are the value of y corresponding to $x = 0$? What are the values of y corresponding to $x = 3$? The graph of the equation $x^2 + y^2 - 6x - 4y = 12$ is the circle given below. What is the value of y corresponding to $x = -2$? Using the equation $x^2 + y^2 - 6x - 4y = 12$ we can define two continuous functions on the interval $-2 \le x \le 8$. State a complete definition of each of these functions.

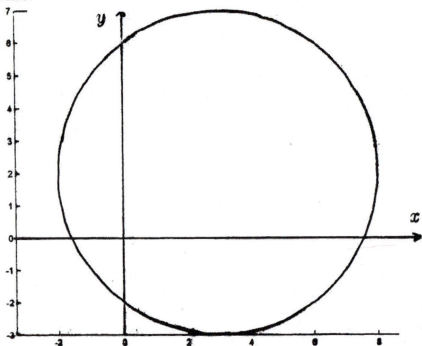

4. How many implicit continuous functions are defined by the equation $x^4 + 16y^4 = 256$. State a complete definition of one of these functions. For what value of x is this function defined?

5. Consider the equation $x = \cos y$. Find all the values of y in the interval $-2\pi \le y \le 2\pi$ corresponding to $x = 0$, $x = \sqrt{2}/2$, and $x = -\sqrt{2}/2$. The variable y is measured in radians. Clearly the equation $x = \cos y$ can be used to define several continuous functions. Give the definition of the function usually called $\arccos x$. State the definition of another continuous function defined by the equation $x = \cos y$ with x as the independent variable.

74

6337 Differentiating Implicit Functions

Suppose we have a function $y = f(x)$ which is defined implicitly using some equation together with one or more side conditions. One question is: how can we find the derivative of such a function? If we are able to solve the equation and get an explicit expression for the function, then we can find the derivative using the usual rules of differentiation. But suppose we have an equation that we cannot solve for y in terms of x, then what do we do? As it turns out finding a formula for the derivative in this case is fairly easy. However, we must spend some time understanding exactly what the formula says once we get the formula. We will explain how to apply these new kind of derivative formulas using some examples. We start with a simple example, one for which we can actually solve for y in terms of x.

Example 1. Let $y = y(x)$ denote the function such that given x the corresponding value of y is the solution of the equation

$$y^2 - 10y - 3x + 16 = 0.$$

Find the derivative of this function at some points.

Solution. What is the value of y corresponding to $x = 0$? Substituting $x = 0$ into the defining equation, we get $y^2 - 10y + 16 = 0$ or $(y - 8)(y - 2) = 0$. There are two values of y. They are 2 and 8. The fact that we have two values of y indicates that the equation $y^2 - 10y - 3x + 16 = 0$ defines two functions. Let us not be concerned for the moment about the fact that there are two functions defined by this equation. Let us just try to find $\frac{dy}{dx}$.

The important fact to notice when finding the derivative is that y is a function of x. In order to differentiate y with respect to x we will need the chain rule. Using the chain rule:

$$\frac{d}{dx}(y^2) = \frac{d}{dy}(y^2)\frac{dy}{dx} = 2y\frac{dy}{dx}.$$

Starting with the equation

$$y^2 - 10y = 3x - 16,$$

let us take the derivative of both sides:

$$\frac{d}{dx}[y^2 - 10y] = \frac{d}{dx}[3x - 16],$$

or

$$\frac{d}{dx}[y^2] - 10\frac{d}{dx}[y] = 3.$$

Using the chain rule as above, we have

$$2y\frac{dy}{dx} - 10\frac{dy}{dx} = 3,$$

$$(2y - 10)\frac{dy}{dx} = 3,$$

or

$$\frac{dy}{dx} = \frac{3}{2y - 10}.$$

This is a formula for $\frac{dy}{dx}$. In our previous work we never saw a formula for $\frac{dy}{dx}$ that involved y. But this is still a good formula. We just need to understand this new kind of formula. We have already discovered that there are two functions of y defined by the equation $y^2 - 10y - 3x + 16 = 0$. In order to have a complete definition of either of these two functions we would need to state the side condition that forms part of the definition. The important thing to realize is that this derivative formula is really two formulas. It is the formula for whichever of the two functions for y we use to replace y. Let $y = f(x)$ denote the first function defined by the equation $y^2 - 10y - 3x + 16 = 0$, the one for which $f(0) = 8$, then the formula for the derivative of that function is

$$f'(x) = \frac{3}{2f(x) - 10}.$$

Substituting values of $x = 0$ into this formula, we get $f'(0) = \frac{3}{2f(0)-10}$. In order to find $f'(0)$, we need to know that $f(0) = 8$. It follows that

$$f'(0) = \frac{3}{2(8) - 10} = \frac{1}{2}.$$

Let $y = g(x)$ denote the second function defined by the equation $y^2 - 10y - 3x + 16 = 0$, the one for which $g(0) = 2$, then replacing y with $g(x)$ we get the formula for $g'(x)$ for all x.

$$g'(x) = \frac{3}{2g(x) - 10}.$$

In order to find $g'(0)$ we must first know that $g(0) = 2$. It follows that

$$g'(0) = \frac{3}{2(2) - 10} = -\frac{1}{2}.$$

We can get a better understanding of what is happening here if we look at the graph of the equation $y^2 - 10y + 25 - 3x + 16 = 0$. We complete the square and get $(y-5)^2 = 3(x+3)$. The graph of this equation is a parabola with its vertex at $(-3, 5)$.

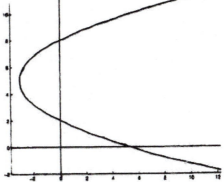

The top half of the parabola is the graph of the first function $y = f(x)$ and the bottom half of the parabola is the graph of the second function $y = g(x)$. Both the functions $f(x)$ and $g(x)$ are only defined for $x \geq -3$. From the graph we can see that $f(0) = 8$ and $g(0) = 2$.

Substituting $x = 9$ into $y^2 - 10y - 3x + 16 = 0$, we get

$$y^2 - 10y - 11 = 0$$
$$(y - 11)(y + 1) = 0.$$

The two values of y corresponding to $x = 9$ are $y = 11$ and $y = -1$. The point on the top half of the parabola is $(9, 11)$. This means that $f(9) = 11$. The derivative (slope of tangent line) for the top function $f(x)$ at $x = 9$ is

$$f'(9) = \frac{3}{2(11) - 10} = \frac{1}{4}.$$

The point on the bottom half of the parabola is $(9, -1)$. This means that $g(9) = -1$. The derivative (slope of the tangent) for the bottom function $g(x)$ at $x = 9$ is

$$g'(9) = \frac{3}{2(-1) - 10} = -\frac{1}{4}.$$

Example 2. Let y be defined as a function of x by the rule that given x the corresponding value of y is the solution of the equation

$$y^2 + 9x^2 - 6y - 36x = 180.$$

Find $\dfrac{dy}{dx}$ for all continuous functions defined by this equation.

Solution. We differentiate both sides of the equation

$$\frac{d}{dx}[y^2 + 9x^2 - 6y - 36x] = \frac{d}{dx}[180]$$

$$\frac{d}{dx}[y^2] + 18x - 6\frac{d}{dx}[y] - 36 = 0.$$

We must keep in mind that y is a function of x. We use the chain rule to find

$$\frac{d}{dx}[y^2] = \frac{d}{dy}[y^2]\frac{dy}{dx} = 2y\frac{dy}{dx}.$$

Using this equality the equation becomes

$$2y\frac{dy}{dx} + 18x - 6\frac{dy}{dx} - 36 = 0$$

$$y\frac{dy}{dx} + 9x - 3\frac{dy}{dx} - 18 = 0$$

$$(y - 3)\frac{dy}{dx} = -9x + 18$$

$$\frac{dy}{dx} = \frac{-9x + 18}{y - 3}.$$

This is a formula for the derivative which is good for any value of x and y and is good for any continuous function defined by the equation. Suppose $x = 6$. The corresponding values of y are the solutions of the equation

$$y^2 + 324 - 216 - 6y - 180 = 0$$

$$y^2 - 6y - 72 = 0$$

$$(y - 12)(y + 6) = 0.$$

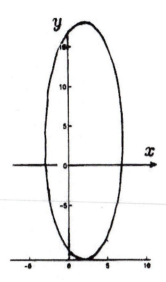

The values of y corresponding to $x = 6$ are $y = 12$ and $y = -6$. The equation $y^2 + 9x^2 - 6y - 36x = 180$ defines two functions y of the independent variable x. A point on the first curve is $(6, 12)$. Call this function $y = f(x)$, then $f(6) = 12$. For this function

$$f'(x) = \frac{-9x + 18}{f(x) - 3}.$$

It follows that

$$f'(6) = \frac{-9(6) + 18}{f(6) - 3} = \frac{-36}{12 - 3} = -4.$$

A point on the second curve is $(6, -6)$. Let $y = g(x)$ denote the function whose graph is the second curve. For the function $g(x)$, we have

$$g'(x) = \frac{-9x + 18}{g(x) - 3}.$$

It follows that

$$g'(6) = \frac{-9(6) + 18}{g(6) - 3} = \frac{-36}{-6 - 3} = 4.$$

Suppose $x = 7$. The corresponding values of y are the solutions of the equation

$$y^2 + 9(7)^2 - 6y - 36(7) = 180$$
$$y^2 - 6y + 9 = 0$$
$$(y - 3)(y - 3) = 0.$$

The only solution is $y = 3$. There is only one point on the curve with x coordinate $x = 7$, namely $(7, 3)$. Substituting $y = 3$ into the formula for the derivative, we get

$$\frac{dy}{dx} = \frac{-9x + 18}{y - 3} = \frac{-27 + 18}{3 - 3}.$$

This shows that $\frac{dy}{dx}$ is not defined when $x = 7$. The graph of the equation $y^2 + 9x^2 - 36x - 6y = 180$ makes the situation clear. This curve is an ellipse with center $(2,3)$. The graph of the first function $y = f(x)$ is the top half of the ellipse. For this function we have already shown that $f(6) = 12$ and $f'(6) = -4$. This function is only defined for $-3 \leq x \leq 7$. The graph of the second function is the bottom half of the ellipse. We have already shown that for this function $g(6) = -6$ and $g'(6) = 4$. This function is only defined for $-3 \leq x \leq 7$. We can find either $f'(x)$ or $g'(x)$ for any value of x such that $-3 < x < 7$. If we look back at the definition of derivative we see that in order to find $f'(7)$ we must be able to find values of $f(x)$ when x is such that $x > 7$ as well as when $x < 7$. But this $f(x)$ is not defined for $x > 7$ and so we can not find $f(x)$ for $x > 7$. This means that we cannot find $f'(7)$ because it is not defined. By looking at the graph we see that the tangent line to the ellipse is vertical at the point where $x = 7$.

Example 3. Let y be defined as a function of x by the rule that given x the corresponding value of y is a solution of the equation

$$x^3 y^2 + 5x^2 = 8xy - 3x + 50.$$

Find $\dfrac{dy}{dx}$ at the point $(2, 3)$.

Solution. First, we are told that $x = 2$ and $y = 3$ is a solution of this equation. Maybe we should check this. We are not told how many functions are defined by this equation. In fact, there are at least two functions defined by this equation. The statement of the problem indicates that we are to consider the function $y = f(x)$ and that $f(x)$ is defined by two properties. First, for each x the corresponding value of y is a solution of $x^3 y^2 + 5x^2 = 8xy - 3x + 50$. Second, one point on the curve $y = f(x)$ is $(2, 3)$. These two properties completely define a function. We can if need be start with these two properties and find out anything else we need to know about the function. Although we have no clear idea what a graph of this function might look like we do know how to find a formula for its derivative. Our problem is to find the derivative of this function at the point $(2, 3)$. We are to find the derivative at $x = 2$ and are lucky to have been given the corresponding y value, namely, $y = 3$.

We start by differentiating both sides of the equation with respect to x.

$$\frac{d}{dx}[x^3y^2 + 5x^2] = \frac{d}{dx}[8xy - 3x + 50]$$

$$\frac{d}{dx}[x^3y^2] + 5\frac{d}{dx}[x^2] = 8\frac{d}{dx}[xy] - 3\frac{d}{dx}[x].$$

In order to find $\frac{d}{dx}[x^3y^2]$, we use the product rule

$$\frac{d}{dx}[x^3y^2] = x^3\frac{d}{dx}[y^2] + y^2\frac{d}{dx}[x^3].$$

We already used the chain rule to find that

$$\frac{d}{dx}[y^2] = \frac{d}{dy}[y^2]\frac{dy}{dx} = 2y\frac{dy}{dx}.$$

Using this we get

$$\frac{d}{dx}[x^3y^2] = 2x^3y\frac{dy}{dx} + y^2(3x^2).$$

We find $\frac{d}{dx}(xy)$ using the product rule

$$\frac{d}{dx}(xy) = x\frac{d}{dx}(y) + y\frac{d}{dx}(x) = x\frac{dy}{dx} + y.$$

Substituting back into the equation using these two equalities, we get

$$2x^3y\frac{dy}{dx} + 3x^2y^2 + 10x = 8x\frac{dy}{dx} + 8y - 3$$

$$(3x^3y - 8x)\frac{dy}{dx} = -3x^2y^2 - 10x + 8y - 3$$

$$\frac{dy}{dx} = \frac{-3x^2y^2 - 10x + 8y - 3}{3x^3y - 8x}.$$

This is a formula for $\frac{dy}{dx}$. This formula involves both x and y as do most formulas for $\frac{dy}{dx}$ that we obtain using implicit differentiation.

Even when the equation defines y as more than one function of x this formula is good. We need only put in the correct y values for the function we want. For this problem we do not have to find the y value, it is given. Substituting $x = 2$ and $y = 3$ into the formula, we get

$$\frac{dy}{dx} = \frac{-3(2)^2(3)^2 - 10(2) + 8(3) - 3}{3(2)^3(3) - 8(2)} = -\frac{107}{56}.$$

There are actually two values of y corresponding to $x = 2$. This means that the equation defines another function, but we are not concerned with this other function. These functions are not defined for $x = 0$ or the large positive values of x. For example, this equation does not define a function for $x = 10$ and so we can not use the derivative formula to find its derivative when $x = 10$.

Exercises

1. For a given value of x suppose that the corresponding value of y is the solution of the equation $y^2 - 6y - 2x + 1 = 0$. What are the values of y corresponding to $x = -2$ and to $x = 14$? How many continuous functions does this equation define? Find $\frac{dy}{dx}$ for all the continuous functions defined by this equation at all the points where $x = -2$ and the points where $x = 4$. Use implicit differentiation.

2. Suppose the function $y = f(x)$ is defined by saying: for a given value of x suppose that the corresponding value of y is the solution of the equation $y^2 - 4y + x = 4$. What are the values of y corresponding to $x = -1$? Corresponding to $x = 4$? How many continuous functions are defined by the equation $y^2 - 4y + x = 4$. Find $\frac{dy}{dx}$ for all the functions defined by $y^2 - 4y + x = 4$ at all the points where $x = 4$ and where $x = -17$.

3. Suppose that the function $y = f(x)$ is defined by the statement: for a given value of x suppose that the corresponding value of y is the solution of the equation $x^2 + y^2 - 4x - 6y = 87$. What are the values of y corresponding to $x = -8$ and $x = 8$? The complete graph of this equation is a circle with center $(2, 3)$ and radius 10.

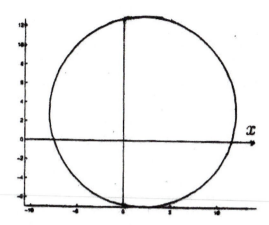

State the side conditions for the two functions defined by $x^2 + y^2 - 4x - 6y = 87$. For what values of x are these two functions defined? You may need to complete the square. Find $\dfrac{dy}{dx}$ for both functions at the point where $x = -8$ and the point where $x = 8$.

4. Find the equation of the tangent line to the curve whose equation is

$$xy^2 + 4y^2 - 3x^3 = 42$$

at the point $(-2, 3)$. Check to see that the given point is really on the curve and then find $\dfrac{dy}{dx}$ at the point.

5. If y is defined as a continuous function of x by the equation $x^3 y + y^3 = 5x + 41$, find $\frac{dy}{dx}$ at the point $(2, 3)$.

6. If y is defined as a continuous function of x by the equation $y^4 + 4x^2 y^2 + 4x^4 = y - 3x + 473$, find $\frac{dy}{dx}$ at the point $(-3, 2)$. $\frac{dy}{dx}\big|_{(-3,2)} = 3$.

7. Suppose y is defined as a function of x as follows. Given a value of x the corresponding value of y is a solution of the equation

$$y^3 - 9y^2 + 15y - x + 13 = 0.$$

(a) How many continuous functions are defined?

(b) For exactly what values of x is each function defined?

(c) Find $\frac{dy}{dx}$ at the points $(13, 0)$, $(4, 3)$, and $(-5, 6)$.

(d) What is the value of $\frac{dy}{dx}$ at $(20, 1)$?

Below is a graph of the equation.

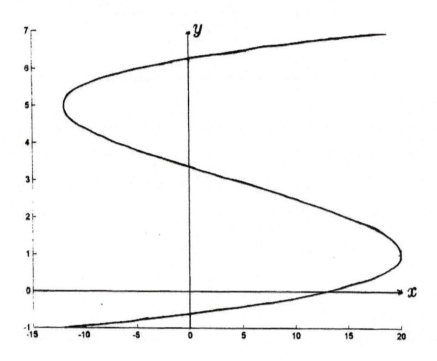

6339 Derivatives of Inverse Functions

We have defined several functions by saying that the new function is the inverse of a function that is already known. That is, the new function is defined implicitly. We define the function $y = \log_e x = \ln x$ to be the function such that given a value of x with $x > 0$ the corresponding value of y is the solution of the equation $x = e^y$. When we require that $x > 0$ there is only one solution to the equation $x = e^y$. This means that we do not need an extra side condition in order to define $y = \ln x$. The problem we want to consider now is to find $\frac{d}{dx}(\ln x)$. Let us differentiate both sides of the equation $x = e^y$ using implicit differentiation.

$$\frac{d}{dx}(x) = \frac{d}{dx}(e^y) = \frac{d}{dy}(e^y)\frac{dy}{dx} = e^y\frac{dy}{dx}$$

$$1 = e^y\frac{dy}{dx} \quad \text{or} \quad \frac{dy}{dx} = \frac{1}{e^y}.$$

The equation is $e^y = x$ and so $\frac{dy}{dx} = \frac{1}{x}$. We have shown that

$$\frac{d}{dx}(\ln x) = \frac{1}{x}, \quad \text{for } x > 0.$$

We sometimes discuss the function

$$\log_{10} x = \frac{\ln x}{\ln 10}.$$

Since $\ln(10)$ is a constant,

$$\frac{d}{dx}(\log_{10} x) = \frac{1}{x(\ln 10)}.$$

The function $\log_{10}(x)$ is not as nice as the function $\ln(x)$ when we start to find derivatives. However, the function $\log_{10} x$ is very interesting when doing calculations. The function $\log_{10}(x)$ also obeys the laws

$$\log_{10}(bc) = \log_{10}(b) + \log_{10}(c)$$
$$\log_{10}(b/c) = \log_{10}(b) - \log_{10}(c)$$
$$\log_{10}(b^x) = x\log_{10}(b).$$

Using these laws we can show for example that

$$\log_{10}(25946) = 4 + \log_{10}(2.5946).$$

We have already discussed the exponential function e^x. This is an opportunity to mention the function a^x where $a > 0$ and $a \neq 1$. As an example, suppose we are given the function $f(x) = 10^x$. Let

$$y = 10^x$$
$$\ln y = \ln(10^x) = x\ln(10)$$
$$y = e^{(\ln 10)x}$$
$$10^x = e^{(\ln 10)x}$$

We then find the derivative.

$$f'(x) = \frac{d}{dx}[10^x] = \frac{d}{dx}[e^{(\ln 10)x}] = (\ln 10)e^{(\ln 10)x} = (\ln 10)(10^x).$$

Anytime we see the function a^x we should convert it using the formula $a^x = e^{(\ln a)x}$.

Next let us consider the function y defined as an implicit function of x by the equation $x = \sin y$. Some values of y corresponding to $x = 0$ are $y = 0$, $y = \pi$, $y = 2\pi$, and $y = -\pi$. It is clear that y is defined as many different functions of x by the equation $x = \sin y$. We need a side condition in order to restrict our consideration to a single function. Let us make the restriction that $-\pi/2 \leq y \leq \pi/2$. When this particular restriction is made the solution of $x = \sin y$ is called arcsin x. Recall that $|\sin y| \leq 1$. This means that in order for the equation $x = \sin y$ to have a solution we must choose x such that $|x| \leq 1$.

Definition 1. The function $y = \arcsin x$ defined for $|x| \leq 1$ is the solution of the equation $x = \sin y$ such that $-\pi/2 \leq y \leq \pi/2$. The function $y = \arcsin x$ is not defined for $|x| > 1$.

We are discussing the function $y = f(x)$ defined as the solution of $x = \sin y$ subject to the side conditions $-\pi/2 \leq y \leq \pi/2$ and $|x| \leq 1$. For this function there is exactly one allowed value of y corresponding to each allowed value of x. When this is the case the function $\arcsin(x)$ is called the inverse of the function $\sin x$. When we use the notation $f(x) = \sin x$ we sometimes use the notation $f^{-1}(x)$ for the inverse function $\arcsin(x)$.

In order to find the derivative of the function $\arcsin x$ we differentiate the equation $x = \sin y$ using implicit differentiation.

$$\frac{d}{dx}(x) = \frac{d}{dx}(\sin y) = \frac{d}{dy}(\sin y)\frac{dy}{dx}$$

$$1 = \cos y \frac{dy}{dx}$$

$$\frac{dy}{dx} = \frac{1}{\cos y}.$$

Given the values of $\cos y$ that satisfies the above restrictions on y, the formula $\frac{1}{\cos y}$ is a correct formula for the derivative of $\arcsin x$. However, we would very much like to have a formula that gives the derivative in terms of x. Recall that

$$\cos^2 y + \sin^2 y = 1$$
$$\cos^2 y = 1 - \sin^2 y$$
$$\cos y = \sqrt{1 - \sin^2 y} \text{ or } \cos y = -\sqrt{1 - \sin^2 y}.$$

Which of these two equations is correct? Do we use the plus square root or do we use the minus square root? For some values of y the first equation is correct while for other values of y the second equation is correct. When $y = \pi/3$, we have $\cos(\pi/3) = 1/2$ and $\sin(\pi/3) = \sqrt{3}/2$. Since $1/2 = \sqrt{1 - (3/4)}$, we see that the first equation is correct when $y = \pi/3$. When $y = 2\pi/3$, we have $\cos(2\pi/3) = -1/2$ and $\sin(2\pi/3) = \sqrt{3}/2$. Since $(-1/2) = -\sqrt{1 - (3/4)}$, we see that the second equation is correct when $y = 2\pi/3$. But recall that we are only considering values of y such that $-\pi/2 \le y \le \pi/2$. For these values of y which is the correct equation? For $-\pi/2 \le y \le \pi/2$ we know that $\cos y \ge 0$. It follows that we must take the plus square root, that is,

$$\cos y = \sqrt{1 - \sin^2 y} \text{ for } -\pi/2 \le y \le \pi/2.$$

Therefore, since $-\pi/2 \le y \le \pi/2$,

$$\frac{dy}{dx} = \frac{1}{\sqrt{1 - \sin^2 y}}.$$

Our equation is $x = \sin y$ and so it follows that

$$\frac{dy}{dx} = \frac{1}{\sqrt{1 - x^2}}.$$

We have shown that

$$\frac{d}{dx}(\arcsin x) = \frac{1}{\sqrt{1 - x^2}}, \text{ for } -1 < x < 1.$$

For what values of x is this derivative formula correct? First, we know that $y = \arcsin(x)$ is only defined for $-1 \le x \le 1$. There is another small difficulty connected with where this derivative formula is correct. This difficulty does not show itself when we use implicit differentiation to derive the derivative formula. But it is true that this formula does not give the value of the derivative of $\arcsin(x)$ when $x = -1$ and when $x = 1$. We actually know from the beginning of the discussion that the derivative of $\arcsin(x)$ does not exist when $x = -1$ and $x = 1$. Let us consider the case of $x = 1$. Recall that the definition of derivative involved a two sided limit. If we look back at the definition of derivative we see that in order for the derivative to exist at $x = 1$, the function $\arcsin x$ must be defined for some values of x such that $x > 1$ as well as for $x < 1$. The function $\arcsin x$ is only defined for $-1 \le x \le 1$. We can not find a two sided limit involving $\arcsin(x)$ at $x = 1$. At this point, we have not defined the derivative in the case where we only have a one sided limit. Recall that some of the functions that we considered before that were defined implicitly also did not have a derivative at such end points. Therefore, from the start of the discussion we should expect that the derivative of $\arcsin(x)$ may not exist at $x = -1$ and $x = 1$.

Definition 2. The function $y = \arctan x$ is defined as follows. Given x suppose that the corresponding value of y is the solution of the equation $x = \tan y$ such that $-\pi/2 \le y \le \pi/2$, then $y = \arctan x$.

Note that the equation $x = \tan y$ has a solution for all real numbers x. The following derivative formula is proved in a manner similar to the way we prove the formula for the derivative of arcsin x.

$$\frac{d}{dx}(\arctan x) = \frac{1}{1 + x^2}.$$

If the functions arc $\cos(x)$ and arc $\cot(x)$ must be considered we can rely on the well known trigonometric identities

$$\text{arc} \cos(x) = \frac{\pi}{2} - \text{arc} \sin(x) \text{ for } -1 \le x \le 1$$

$$\text{arc} \cot(x) = \frac{\pi}{2} - \text{arc} \tan(x).$$

We hardly ever use the function arc $\sec(x)$ and so probably should skip finding the derivative of arc $\sec(x)$. However, we will discuss the formula for the derivative of arc $\sec(x)$ because there is a small controversy about how to define the function arc $\sec(x)$. The function $y = \text{arc} \sec(x)$ is defined by the statement: suppose that for a given value of x the corresponding value of y is the solution of the equation $x = \sec y$ such that a side condition is satisfied, then $y = \text{arc} \sec(x)$. There are three side conditions that are used. The most logical side condition is to say

"such that $0 \le y < \pi/2$ or $\pi/2 < y \le \pi$".

The other two side conditions that are also used are:

"such that $0 \le y < \pi/2$ or $-\pi \le y < -\pi/2$"

"such that $0 \le y < \pi/2$ or $\pi \le y < 3\pi/2//$.

The second and third conditions are essentially equivalent. The advantages of the first side condition is that it gives the smallest value of y, it makes arc $\sec(x)$ an increasing function, and it satisfies the identity arc $\sec(x) = $ arc $\cos(1/x)$ for $|x| \ge 1$. However, as we shall see this choice of side condition does have a disadvantage. Recall that $|\sec y| \ge 1$. In order to solve the equation $x = \sec y$ the value of x must be such that $|x| \ge 1$. Also recall that $\cos(-\pi/2) = 0$ and $\cos(\pi/2) = 0$, and so $\sec(-\pi/2)$ and $\sec(\pi/2)$ are not defined. There is no value of x such that the corresponding value of y is $-\pi/2$ or $\pi/2$. The function arc $\sec(x)$ is really a function in two parts. Suppose $x \geq 1$, then for all three choices $y = \text{arc}\sec(x)$ is chosen such that $0 \leq y < \pi/2$. Suppose $x \leq -1$, then there are three choices for $y = \text{arc}\sec(x)$. We can choose y such that either $\pi/2 < y \leq \pi$ or $-\pi \leq y < -\pi/2$, or $\pi \leq y < 3\pi/2$. When using the function $\text{arc}\sec(x)$ in a practical problem we either consider $x \geq 1$ only or $x \leq -1$ only.

Suppose the function $y = \text{arc}\sec(x)$ is defined by the statement: suppose for a given value of x the corresponding value of y is the solution of the equation $x = \sec y$ such that $0 \leq y < \pi/2$ or $\pi/2 < y \leq \pi$, then

$$\frac{d}{dx}[\text{arc}\sec(x)] = \frac{1}{|x|\sqrt{x^2 - 1}} \text{ for either } x < -1 \text{ or } x > 1.$$

Suppose the function $y = \text{arc}\sec(x)$ is defined by the statement: suppose for a given value of x the corresponding value of y is the solution of the equation $x = \sec y$ such that $0 \leq y < \pi/2$ or $\pi \leq y < 3\pi/2$, then

$$\frac{d}{dx}[\text{arc}\sec(x)] = \frac{1}{x\sqrt{x^2 - 1}} \text{ for either } x < -1 \text{ or } x > 1.$$

Note that $\sqrt{x^2 - 1}$ is not defined for $|x| < 1$. The disadvantage of using the first method of defining $\text{arc}\sec(x)$ is that the derivative contains $|x|$ where the second method contains x. These two expressions are different for $x < -1$. Even though the function $y = \text{arc}\sec(x)$ is defined for $x = -1$ and $x = 1$ the derivative of $\text{arc}\sec(x)$ is not defined for $x = -1$ and also not for $x = 1$.

We have found the following derivative formulas of the inverse trigonometric functions.

$$\frac{d}{dx}[\arcsin(x)] = \frac{1}{\sqrt{1 - x^2}} \quad \frac{d}{dx}[\arctan(x)] = \frac{1}{1 + x^2}$$

$$\frac{d}{dx}[\text{arcsec }(x)] = \frac{1}{|x|\sqrt{x^2 - 1}}$$

Note that $\frac{d}{dx}[\text{arcsec}(x)] = \frac{1}{x\sqrt{x^2-1}}$ when $x > 1$. We could have found the derivative formulas

$$\frac{d}{dx}[\text{arcsinh }(x)] = \frac{1}{\sqrt{1 + x^2}} \quad \frac{d}{dx}[\text{arctanh }(x)] = \frac{1}{1 - x^2}$$

using $x = \sinh(y)$ and $x = \tanh(y)$.

Example 1. Let $y = x^2 \arctan(5x)$, find $\dfrac{dy}{dx}$.

Solution. This function is expressed as a product. We start with the product rule

$$\frac{dy}{dx} = x^2 \frac{d}{dx}[\arctan(5x)] + [\arctan(5x)]\frac{d}{dx}(x^2).$$

We find $\dfrac{d}{dx}[\arctan(5x)]$ using the chain rule. Let $u = 5x$, then

$$\frac{d}{dx}[\arctan u] = \frac{d}{du}[\arctan u]\frac{du}{dx} = \frac{1}{1+u^2} \quad (5)$$

$$\frac{d}{dx}[\arctan(5x)] = \frac{5}{1+25x^2}.$$

Substituting using this equality, we get

$$\frac{dy}{dx} = \frac{5x^2}{1+25x^2} + 2x[\arctan(5x)].$$

Example 2. Let $y = \left[\sqrt{1-x^4}\right][\arcsin(x^2)]$, find $\dfrac{dy}{dx}$.

Solution. We first apply the product rule.

$$\frac{dy}{dx} = \sqrt{1-x^4}\frac{d}{dx}[\arcsin(x^2)] + [\arcsin(x^2)]\frac{d}{dx}[\sqrt{1-x^4}].$$

We find $\frac{d}{dx}[\arcsin(x^2)]$ using the chain rule. Let $u = x^2$, then

$$\frac{d}{dx}[\arcsin u] = \frac{d}{du}[\arcsin u]\frac{du}{dx} = \frac{1}{\sqrt{1-u^2}}(2x) = \frac{2x}{\sqrt{1-x^4}}.$$

Find $\frac{d}{dx}[\sqrt{1-x^4}]$ using the chain rule. Let $u = 1 - x^4$, then

$$\frac{d}{dx}[\sqrt{u}] = \frac{d}{dx}[u^{1/2}] = \frac{d}{du}(u^{1/2})\frac{du}{dx} = (1/2)u^{-1/2}\frac{du}{dx}$$

$$= \frac{1}{2\sqrt{u}}(-4x^3) = \frac{-4x^3}{2\sqrt{1-x^4}}$$

Substituting using the last two equalities, we get

$$\frac{dy}{dx} = [\sqrt{1 - x^4}]\frac{2x}{\sqrt{1 - x^4}} + [\arcsin(x^2)]\frac{-4x^3}{2\sqrt{1 - x^4}}$$

$$= 2x - \frac{2x^3}{\sqrt{1 - x^4}}[\arcsin(x^2)].$$

Exercises

Find the derivative of each of the following functions.

1. $F(x) = (\ln x)^2 + \ln(x^2)$ when $x > 0$.

2. $f(x) = \dfrac{1 + \ln x}{1 - \ln x}$ defined for $x > e$.

3. $f(x) = \arctan x + \text{arc}\sec(x)$ defined for $x \geq 1$.

4. $g(x) = x^2 \arcsin(x/5)$ defined for $|x| \leq 5$.

5. $F(x) = \dfrac{\arctan(x/2)}{4 + x^2}$.

6. $f(x) = \sqrt{25 - x^2}\arcsin(x/5)$ defined for $|x| < 5$.

7. $g(x) = \ln(1 + x^2) + 5^x$.

8. $f(x) = \ln(\cos x)$ defined for $-\pi/2 < x < \pi/2$.

9. The function $y = \arctan x$ is defined as follows: For a given x the corresponding value of y is the solution of the equation $x = \tan y$ such that $-\pi/2 \leq y \leq \pi/2$. Show that

$$\frac{d}{dx}(\arctan x) = \frac{1}{1 + x^2}.$$

10. Suppose that the function $y = f(x)$ is defined for $|x| \leq 1$ to be the solution of the equation $x = \sin y$ such that $\pi/2 \leq y \leq 3\pi/2$. Find $f'(x)$.

6341 Velocity and other Rates of Change

Suppose an object moves along a straight line and s denotes the directed distance of the object from the origin (zero point) and suppose $s = f(t)$ where t is time. We call $f(t)$ the position function. We have discussed both the ideas of average velocity and instantaneous velocity. The instantaneous velocity is given by $\frac{ds}{dt} = f'(t)$.

Suppose y is a quantity that depends on time t, that is, $y = f(t)$. If t changes from $t = t_1$ to $t = t_2$, then the change in t is often called the increment of t. The corresponding change in y is $\Delta y = f(t_2) - f(t_1)$. The difference quotient

$$\frac{\Delta y}{\Delta t} = \frac{f(t_2) - f(t_1)}{t_2 - t_1}$$

is called the average rate of change of y with respect to t over the interval $t_1 \le t \le t_2$. If we allow t to approach t_2 we get the instantaneous rate of change

$$\lim_{t \to t_2} \frac{f(t_2) - f(t)}{t_2 - t} = f'(t_2).$$

Example 1. Suppose a roast is taken from the oven when its temperature is 185°F and placed on a table in a room where the temperature is 75°F. Suppose that after 10 minutes the temperature of the roast is 100°F. According to Newton' Law of Cooling the temperature $u(t)$ of the roast at time t is given by

$$u(t) = 75 + 110e^{-0.15t}.$$

Find the rate of change of temperature of the roast at time $t = 3$ minutes.

Solution. The rate of change of temperature at time t is given by $u'(t)$.

$$u'(t) = -16.5e^{-0.15t}$$
$$u'(3) = -16.5e^{-0.45} = -10.5.$$

At time $t = 3$ minutes the temperature is changing at the rate of -10.5 degrees per minute. When using the phrase "rate of change" we most commonly are referring to the rate of change with respect to time. We have some variable, say f, which is dependent on time t. In this case the rate of change of f with respect to time is the derivative of f with respect to t. We have discussed this in the case $s = f(t)$ gives the position of an object on a line and t is time. In this case the rate of change of position with respect to time $\frac{ds}{dt} = f'(t)$ is called the velocity.

We can also discuss other rates of change, we use the phrase "rate of change" when the independent variable is not time.

Example 2. A spherical balloon is being inflated. Find the rate of change of volume, V, with respect to radius, r, when $r = 2$ and $r = 3$.

Solution. The volume of a sphere as a function of the radius is given by $V = \frac{4}{3}\pi r^3$. The rate of change of volume with respect to radius is

$$\frac{dV}{dr} = 4\pi r^2.$$

When $r = 2$, we get $\frac{dV}{dr} = 16\pi$. When $r = 3$, we get $\frac{dV}{dr} = 36\pi$. A small change in radius results in a larger increase in volume when $r = 3$ than does the same change in radius when $r = 2$.

Example 3. The cost C (in dollars) of producing x units of a certain commodity is
$$C(x) = 4,000 + 12x + 0.08x^2.$$

Find the rate of change of cost, C, with respect to x when $x = 120$. (This is called marginal cost).

Solution. The rate of change or marginal cost for any value of x is given by the derivative.
$$C'(x) = 12 + 0.16x.$$
$$C'(120) = 12 + 0.16(120) = 31.2.$$

The marginal cost when $x = 120$ is \$31.20. This means that it costs \$31.20 more to produce 121 units than it cost to produce 120 units approximately. The actual increase in cost is

$$C(121) - C(120) = 12(121 - 120) + 0.08[121^2 - 120^2] = \$31.28$$

Problems

1. A man is standing on a tall ladder and throws a ball into the air. The height of the ball after t seconds is given by

$$h(t) = 200 + 300t - 16t^2.$$

What is the rate of change of height of the ball with respect to time when $t = 5$ seconds? When $t = 12$ seconds?

2. A steel ball is heated to a temperature of $280°$ and placed in a room where the temperature is $80°$. After 6 minutes the temperature of the ball is $148°$. According to Newton's Law of Cooling the temperature $u(t)$ of the ball at any time t is given by

$$u(t) = 80 + 200e^{-0.18t}.$$

What is the rate of change of temperature with respect to time when $t = 8$ minutes?

3. Suppose the cost of production C in dollars of producing x units of a certain commodity is

$$C(x) = 5000 + 15x + 0.09x^2.$$

Find the rate of change of cost C with respect to x when $x = 50$.

4. The frequency of vibrations of a vibrating string is given by

$$f = \frac{1}{2L}\sqrt{\frac{T}{\rho}},$$

where L is the length of the string, T is the tension in the string, and ρ is its linear density. Assume T and ρ are kept constant, find the rate of change of frequency with respect to length.

5. A spherical balloon is being inflated. Find the rate of increase of surface area with respect to radius.

6. If a gas is kept at a constant temperature, then its volume, V depends on the pressure, p. Suppose that the volume of a certain gas at a certain temperature is related to pressure by

$$V = \frac{8}{p},$$

Find the rate of change of volume with respect to pressure. Pressure is usually measured in pounds per square inch or in kilopascals.

7. Consider the graph of the function $y = f(x)$. Suppose that the value of y corresponding to any value of x is given by the cubic

$$y = x^3 - 5x^2 + 12x + 108.$$

Find the rate of change of y with respect to x when $x = 5$. What does this rate of change number have to do with tangent lines?

8. Consider a right circular cylinder. The volume is $V = \pi r^2 h$. Suppose we let the radius vary but keep the height constant at $h = 25$ft. What is the rate of change of volume with respect to radius when the radius is $r = 10$ft?

6343 Higher Derivatives

In our previous discussion of differentiation we have always assumed that we just started with some function $f(x)$ and wanted to find its derivative $f'(x)$. However, it is clear that we could repeat this operation. Once we have found the derivative $f'(x)$ we could also find the derivative of $f'(x)$. This derivative of the derivative is called the second derivative and is denoted by $f''(x)$. Using the Leibniz notation the second derivative of $y = f(x)$ is written as

$$\frac{d^2y}{dx^2}.$$

Example 1. If $f(x) = x^2 \sin x$, find the second derivative $f''(x)$.

Solution. First, using the product rule, we find the first derivative which is

$$f'(x) = 2x(\sin x) + x^2(\cos x).$$

We find the second derivative using the product rule to differentiate each term:

$$f''(x) = 2\sin x + 2x\cos x + 2x\cos x - x^2 \sin x$$
$$= 2\sin x + 4x\cos x - x^2 \sin x.$$

We can even find the third derivative by differentiating the second derivative.

$$f'''(x) = 2\cos x + 4\cos x - 4x\sin x - 2x\sin x - x^2 \cos x$$
$$= 6\cos x - 6x\sin x - x^2 \cos x.$$

Suppose $s = s(t)$ denotes the position of an object that is moving along a straight line, then the first derivative is the velocity of the object, $v(t) = s'(t)$. The instantaneous rate of change of velocity with respect to time is called the acceleration, $a(t)$, of the object. The acceleration function is the derivative of the velocity function. Thus the acceleration is the second derivative of the position function.

Example 2. Suppose that the position, s, of a particle moving back and forth on a straight line is given at time t by

$$s = 2t^3 - 8t^2 + 11t.$$

Find the position, velocity, and acceleration of this particle when $t = 2$.

Solution. Assume t is measured in seconds and s is measured in feet. The position of the object at $t = 2$ is

$$s(2) = 2(2)^3 - 8(2)^2 + 11(2) = 6.$$

The position of the object is 6 feet to the right of the zero mark. This assumes that we indicate position to the right of the zero mark using positive numbers. In order to find the velocity when $t = 2$, we first find an expression for velocity at any time t.

$$s'(t) = 6t^2 - 16t + 11$$
$$s'(2) = 6(2)^2 - 16(2) + 11 = 3.$$

The velocity is 3 ft/sec in the positive direction (to the right).

The acceleration at any time t is given by

$$s''(t) = 12t - 16.$$
$$s''(2) = 12(2) - 16 = 8.$$

The acceleration is 8 ft/sec^2 in the positive direction at time $t = 2$.

Example 3. Suppose y is defined implicitly as a function of x by the equation

$$x^4 + y^4 = 32.$$

Find the second derivative $\dfrac{d^2y}{dx^2}$ at $x = 2$.

Solution. We find the first derivative using implicit differentiation. We have used implicit differentiation before. Recall that this involves using the chain rule. Using the chain rule we find that

$$\frac{d}{dx}(y^4) = \frac{d}{dy}(y^4)\frac{dy}{dx} = 4y^3\frac{dy}{dx}.$$

Use this result when we differentiate both sides of the given equation with respect to x.

$$\frac{d}{dx}(x^4) + \frac{d}{dx}(y^4) = \frac{d}{dx}(16).$$

$$4x^3 + 4y^3 \frac{dy}{dx} = 0.$$

We could now solve for $\frac{dy}{dx}$ in terms of x and y. However, if we want to find the second derivative it is best not to solve for $\frac{dy}{dx}$ at this point. We need to differentiate this equation again. The most difficult part of this differentiation is to find

$$\frac{d}{dx}\left[y^3 \frac{dy}{dx}\right].$$

In order to find this derivative we note that this is a product and so we must start by using the product rule.

$$\frac{d}{dx}\left[y^3 \frac{dy}{dx}\right] = y^3 \frac{d}{dx}\left[\frac{dy}{dx}\right] + \left(\frac{dy}{dx}\right)\frac{d}{dx}\left[y^3\right].$$

The chain rule says that $\frac{d}{dx}(y^3) = 3y^2 \frac{dy}{dx}$. Using this fact we have

$$\frac{d}{dx}\left[y^3 \frac{dy}{dx}\right] = y^3 \frac{d^2y}{dx^2} + \left(\frac{dy}{dx}\right)\left[3y^2 \frac{dy}{dx}\right]$$

$$= y^3 \frac{d^2y}{dx^2} + 3y^2 \left(\frac{dy}{dx}\right)^2.$$

Now let us take the derivative of both sides of the equation

$$x^3 + y^3 \frac{dy}{dx} = 0.$$

The result is

$$\frac{d}{dx}\left[x^3\right] + \frac{d}{dx}\left[y^3 \frac{dy}{dx}\right] = \frac{d}{dx}[0]$$

$$3x^2 + y^3 \frac{d^2y}{dx^2} + 3y^2 \left(\frac{dy}{dx}\right)^2 = 0.$$

Solving for the second derivative, we get

$$\frac{d^2y}{dx^2} = \frac{-3x^2 - 3y^2\left(\frac{dy}{dx}\right)^2}{y^3}.$$

We would like to have this formula in terms of x and y. From our work above

$$\frac{dy}{dx} = -\frac{x^3}{y^3}.$$

Substituting, we get

$$\frac{d^2y}{dx^2} = \frac{-3x^2 - 3y^2(-x^3/y^3)^2}{y^3}$$

$$= \frac{-3x^2y^4 - 3x^6}{y^7}$$

This is a formula for $\frac{d^2y}{dx^2}$ in terms of x and y. We want to find $\frac{d^2y}{dx^2}$ at the point where $x = 2$. If we look at the graph of the equation $x^4 + y^4 = 32$, we see that there are two points with x coordinate 2. If we substitute $x = 2$ into $x^4 + y^4 = 32$, we are able to solve for y.

$$16 + y^4 = 32$$
$$y = 2, -2.$$

There are two points namely $(2, 2)$ and $(2, -2)$ on the graph with x coordinate 2. This implies two functions. Which function should we use? The statement of the problem is not clear. The equation $x^4 + y^4 = 32$ defines y as two different functions of x. The values for $\frac{d^2y}{dx^2}$ for these two functions at $x = 2$ are different. We need an extra side condition to be clear which function we are supposed to use. By far the easiest way to be clear about which function we are suppose to use is to give both coordinates of the point on the graph of the function we are using. This means that when we are asked to find $\frac{d^2y}{dx^2}$ for a function that is defined implicitly, we give both the x and y values of the function at the point in question. We should rewrite the question in this example as: Find $\dfrac{d^2y}{dx^2}$ at the point $(2, -2)$. Substituting $x = 2$, $y = -2$ into the formula for $\dfrac{d^2y}{dx^2}$ given above, we get

$$\frac{d^2y}{dx^2}\bigg|_{(2,-2)} = \frac{-3(2)^2(-2)^4 - 3(2)^6}{(-2)^7} = 3.$$

Exercises

1. Find the second derivative $f''(x)$ for

    (a) $f(x) = \arctan(2x)$

    (b) $f(x) = x^4 + 5x^3 + 7x + 12$

    (c) $f(x) = \ln(1 + x^2) + x \arctan x$.

2. Suppose $s(t)$ denotes the position at time t of an object moving along a straight line. Find the velocity and acceleration of the object at the time when $t = 2$.

    (a) $s(t) = t^3 - 5t^2 + 7t$

    (b) $s(t) = 4t^2 + \frac{8}{t+2}$.

3. Suppose y is defined implicitly as a function of x by the equation

$$2x^2 + y^2 - 8x + 6y = 19.$$

Find a formula for $\frac{dy}{dx}$ and $\frac{d^2y}{dx^2} =$ in general. Then find the numerical value of $\frac{dy}{dx}$ and $\frac{d^2y}{dx^2}$ at the point $(2, 3)$. $\frac{dy}{dx} = 0$ and $\frac{d^2y}{dx^2} = -\frac{1}{3}$.

6347 Hyperbolic Functions

Certain functions collectively called hyperbolic functions arise a lot in applications. These functions have the same relationship to the hyperbola that the trigonometric functions have to the circle. However, the hyperbolic functions are more easily defined in terms of exponential functions instead of relating them to the graph of a hyperbola.

Definition of the hyperbolic functions.

$$\sinh x = \frac{e^x - e^{-x}}{2} \qquad\qquad \operatorname{csch} x = \frac{1}{\sinh x}$$

$$\cosh x = \frac{e^x + e^{-x}}{2} \qquad\qquad \operatorname{sech} x = \frac{1}{\cosh x}$$

$$\tanh x = \frac{\sinh x}{\cosh x} \qquad\qquad \coth x = \frac{\cosh x}{\sinh x}.$$

Note that only $\sinh x$ and $\cosh x$ are really new functions. We will only discuss the first three functions $\sinh x$, $\cosh x$, and $\tanh x$ in any detail.

By just looking at the definitions we see that $\sinh(0) = 0$, $\cosh(0) = 1$, $\tanh(0) = 0$. Also $\sinh(-x) = -\sinh(x)$, $\cosh(-x) = \cosh(x)$ and $\tanh(-x) = -\tanh(x)$. The most famous application of hyperbolic functions is to describe the shape of a hanging cable. If a flexable cable is suspended between two points of the same height, then the shape of the cable is given by $y = c + a\cosh(x/a)$. This would be the shape of an electric wire hanging between two poles.

Just as with circular functions there are many identities connecting the hyperbolic functions. The two fundamental hyperbolic identities involve the hyperbolic sine and cosine and are the following:

(I) $\sinh(x + y) = \sinh(x)\cosh(y) + \cosh(x)\sinh(y)$

(II) $\cosh(x + y) = \cosh(x)\cosh(y) + \sinh(x)\sinh(y)$.

Proof of (I). In order to prove this identity we must express $\sinh(x)$ and $\cosh(y)$ using the definitions.

$$\sinh(x)\cosh(y) + \cosh(x)\sinh(y)$$

$$= \left(\frac{e^x - e^{-x}}{2}\right)\left(\frac{e^y + e^{-y}}{2}\right) + \left(\frac{e^x + e^{-x}}{2}\right)\left(\frac{e^y - e^{-y}}{2}\right)$$

$$= \frac{e^{x+y} + e^{x-y} - e^{-x+y} - e^{-x-y}}{4} + \frac{e^{x+y} - e^{x-y} + e^{-x+y} - e^{-x-y}}{4}$$

$$= \frac{2e^{x+y} - 2e^{-x-y}}{4} = \frac{e^{x+y} - e^{-(x+y)}}{2} = \sinh(x+y)$$

We obtain another identity by replacing y with $-y$ in identity (I).

$$\sinh(x - y) = \sinh(x)\cosh(-y) + \cosh(x)\sinh(-y).$$

Since $\cosh(-y) = \cosh(y)$ and $\sinh(-y) = -\sinh(y)$ we can rewrite this as

$$\sinh(x - y) = \sinh(x)\cosh(y) - \cosh(x)\sinh(y).$$

We can obtain new identities by making substitutions into already known identities. Let us replace y with x in identity (II).

$$\cosh(x + x) = \cosh(x)\cosh(x) + \sinh(x)\sinh(x)$$
$$\cosh(2x) = \cosh^2(x) + \sinh^2(x).$$

Let us replace y with $(-x)$ in identity (II).

$$\cosh(x - x) = \cosh(x)\cosh(-x) + \sinh(x)\sinh(-x)$$
$$\cosh(0) = \cosh(x)\cosh(x) - \sinh(x)\sinh(x)$$
$$1 = \cosh^2(x) - \sinh^2(x).$$

As we can see from these identities the hyperbolic identities are almost like trigonometric identities but are never exactly the same. Just as with trigonometric identities we can now find hundreds of hyperbolic identities. However, we will not do that, thank goodness.

Now let us turn to the problem of finding the derivatives of the hyperbolic functions. In order to find the derivatives of the hyperbolic functions, we use their definitions. Let us find the derivative of $\sinh(x)$.

$$\sinh(x) = (1/2)(e^x - e^{-x}).$$

$$\frac{d}{dx}[\sinh(x)] = \frac{1}{2}\frac{d}{dx}[e^x - e^{-x}] = \frac{1}{2}[e^x + e^{-x}] = \cosh(x).$$

$$\cosh(x) = (1/2)(e^x + e^{-x}).$$

$$\frac{d}{dx}[\cosh(x)] = \frac{1}{2}\frac{d}{dx}[e^x + e^{-x}] = \frac{1}{2}[e^x - e^{-x}] = \sinh(x).$$

Using the quotient rule we find that

$$\frac{d}{dx}[\tanh(x)] = \operatorname{sech}^2(x).$$

The Inverse Hyperbolic Functions

The function inverse hyperbolic sinh is denoted by $y = \operatorname{arcsinh}(x)$ and is defined as the solution of the equation $x = \sinh(y)$. Note that we are able to solve this equation given any real number x. A sketch of the graph of $\sinh(y)$ follows.

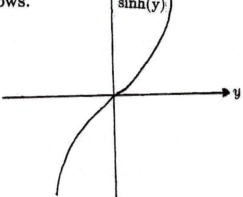

For each value of y there is only one corresponding value of $\sinh(y)$. This means that we do not need an extra side condition in order to define the function arcsinh(x).

Definition. For a given value of x the value of $y = \operatorname{arcsinh}(x)$ is the unique solution of the equation $x = \sinh(y)$.

In order to find the derivative of arc sinh(x) we need to use implicit differentiation. If $y = \operatorname{arcsinh}(x)$, then

$$x = \sinh(y).$$

Differentiate both sides of this equation with respect to x, we get

$$\frac{d}{dx}[x] = \frac{d}{dx}[\sinh(y)]$$

$$1 = \frac{d}{dy}[\sinh(y)]\frac{dy}{dx}$$

$$1 = \cosh(y)\frac{dy}{dx}$$

$$\frac{dy}{dx} = \frac{1}{\cosh(y)}.$$

We would like a formula for $\frac{dy}{dx}$ that only involves x. Recall the identity

$$\cosh^2(y) = 1 + \sinh^2(y).$$

This tells us that either

$$\cosh(y) = +\sqrt{1 + \sinh^2(x)} \quad \text{or} \quad \cosh(y) = -\sqrt{1 + \sinh^2(y)}.$$

For all y we know that $\cosh(y) \geq 1 > 0$. This means that $\cosh(y)$ can never be given by the negative square root. Therefore, for all real numbers y

$$\cosh(y) = \sqrt{1 + \sinh^2(y)}.$$

Substituting for $\cosh(y)$, we get

$$\frac{dy}{dx} = \frac{1}{\sqrt{1 + \sinh^2(y)}}$$

But $\sinh y = x$. This says for all real numbers x that

$$\frac{d}{dx}[\text{arcsinh}(x)] = \frac{1}{\sqrt{1 + x^2}}.$$

Next, let us look at the problem of defining the function which is the inverse of $\cosh(x)$. We denote this function by $\text{arccosh}(x)$. First, of course, we want to say that given x the value of y such that $y = \text{arccosh}(x)$ is a solution of the equation $x = \cosh(y)$. A sketch of the graph of $\cosh(y)$ is

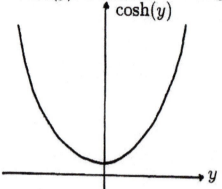

First, note that $\cosh y \geq 1$ so that $x \geq 1$, that is, $\text{arccosh}(x)$ is only defined for $x \geq 1$. Given a value of x there are two values of y such that $\cosh y$ equals this value of x. Since for a given x there are two values of y which are solutions of the equation $x = \cosh(y)$, we need a side condition in order to define the function $\text{arccosh}(x)$. The usual side condition is $\text{arccosh}(x) \geq 0$. However, in the case of the function $\text{arccosh}(x)$ we have an easy way to define $\text{arccosh}(x)$ without having to always state the "side condition". If we solve the equation $x = \cosh(y)$ for y, we get the two solutions

$$y = \ln(x + \sqrt{x^2 - 1}) \text{ and } y = \ln(x - \sqrt{x^2 - 1}).$$

The expression $\ln(x + \sqrt{x^2 - 1})$ is clearly greater than or equal to zero while the expression $\ln(x - \sqrt{x^2 - 1})$ is clearly less than or equal to zero. Note that for $x \geq 1$, it is true that $x + \sqrt{x^2 - 1} \geq 1$ and it is true that $x - \sqrt{x^2 - 1} \leq 1$. Recall we want to say $\text{arccosh}(x) \geq 0$. This means that we can express the function we wish to define as $\text{arccosh}(x)$ explicitly as:

Definition. $\text{arccosh}(x) = \ln(x + \sqrt{x^2 - 1})$ for $x \geq 1$.

We can now find the derivative using the usual rules for differentiation as

$$\frac{d}{dx}[\text{arccosh}(x)] = \frac{d}{dx}[\ln(x + \sqrt{x^2 - 1})]$$
$$= \frac{1}{\sqrt{x^2 - 1}}.$$

Of course, we could also define the function $y = \text{arccosh}(x)$ as follows: Given x the corresponding value of y is the solution of $x = \cosh(y)$ such that $y \geq 0$. We define the function $\text{arctanh}(x)$ as follows: For a given value of x such that $-1 \leq x \leq 1$ the number $y = \text{arctanh}(x)$ is the unique solution of the equation $x = \tanh(y)$. We will not discuss the function $\text{arctanh}(x)$ in any serious way except to remark that in applied mathematics people often use the formula

$$\frac{d}{dx}[\text{arctanh}(x)] = \frac{1}{1 - x^2}$$

in ways that surprise you. So if some day you are surprised by this formula just remember that it exists.

Exercises

1. Prove that the following identities are true:

   (a) $\sinh(x) + \cosh(x) = e^x$.

   (b) $\sinh(2x) = 2\sinh(x)\cosh(x)$

   (c) $\sinh(x - y) = \sinh(x)\cosh(y) - \cosh(x)\sinh(y)$

   (d) $\cosh(x - y) = \cosh(x)\cosh(y) - \sinh(x)\sin(y)$

   (e) $[\cosh(x) + \sinh(x)]^2 = \cosh(2x) + \sinh(2x)$.

2. Prove identity (II): $\cosh(x + y) = \cosh(x)\cosh(y) + \sinh(x)\sinh(y)$.

3. Find the derivative of each of the following:

   (a) $f(x) = \sinh(3x) + \cosh(5x)$

   (b) $f(x) = [\sinh(3x)][\cosh(5x)]$

   (c) $f(x) = x^2 \sinh(4x)$

   (d) $f(x) = x^2 + \sinh^3(4x)$

   (e) $f(x) = \text{arcsinh}(5x) + \text{arccosh}(3x)$.

   (f) $f(x) = \text{arccosh}(x^3) + \sinh(x^2 + 5)$.

   (g) $f(x) = \text{arctanh}(x) - \frac{1}{2}\ln\frac{1+x}{1-x}$.

6349 Differentials

The beginning study of calculus is often divided into two parts. The first part is the study of derivatives and the second part is the study of integrals. We are now engaged in the first part. The first part is also often referred to as the study of "Differential Calculus". This means finding differentials. The next course after calculus for most students is usually a course called "Differential Equations". This means equations involving differentials. However, a more accurate name for this course as it is now taught would be "Derivative Equations". The point here is that the study of differentials is an important topic in calculus.

Definition 1. Given the function $f(x)$, let $y = f(x)$. The differential dx of the independent variable x is also an independent variable. This means dx can be any real number. The differential dy of the dependent variable y is defined as

$$dy = f'(x)dx.$$

The differential dy depends on both x and dx.

Example 1. Let $y = f(x) = x^2 + \sin 3x$, find the differential dy.

Solution. Since we already know how to find the derivative, this is an easy problem.

$$f'(x) = 2x + 3\cos 3x$$
$$dy = (2x + 3\cos 3x)dx.$$

The biggest difficulty is finding the differential is to remember to write down this last step. That is, how to write down the answer. We already know how to find the derivative $f'(x)$. We just need to remember that this is not quite the end of the problem. We still need to write down the answer in the proper form, the differential form.

When we stated the definition of derivative we discussed increments. We now want to discuss increments again in connection with our discussion of differentials.

Definition 2. Give the function $f(x)$, let $y = f(x)$. The increment Δx of the independent variable x is also an independent variable. This means that Δx can be any real number. The increment Δy of the dependent variable

y is defined as

$$\Delta y = f(x + \Delta x) - f(x).$$

The increment Δy depends on both x and Δx.

Example 2. Let $y = 3x^2 + 5x$, find Δy.

Solution.

$$f(x + \Delta x) = 3(x + \Delta x)^2 + 5(x + \Delta x)$$
$$= 3x^2 + 6x(\Delta x) + 3(\Delta x)^2 + 5x + 5\Delta x.$$

$$f(x + \Delta x) - f(x) = [3x^2 + 6x(\Delta x) + 3(\Delta x)^2 + 5x + 5\Delta x] - [3x^2 + 5x]$$
$$= 6x(\Delta x) + 3(\Delta x)^2 + 5\Delta x$$

Therefore, the increment Δy is

$$\Delta y = (6x + 5)\Delta x + 3(\Delta x)^2.$$

We often say that Δy is the amount of change in y corresponding to a change of Δx in x. Although the differential of the independent variable dx, and the increment of the independent variable , Δx, can be any real number, we usually think of them as being small numbers. Note that $dy = (6x + 5)dx$. Suppose $dx = \Delta x$, then

$$\Delta y - dy = (6x + 5)\Delta x + 3(\Delta x)^2 - (6x + 5)dx$$
$$= 3(\Delta x)^2 = 3(dx)^2.$$

If $\Delta x = dx$ is a small number, then $(\Delta x)^2 = (dx)^2$ is a much smaller number. If $\Delta x = dx$ is small, then $\Delta y \approx dy$.

The problem of just finding the differential dy or just finding the increment Δy is not a very interesting problem. For this reason we invent another problem involving Δy and dy which seems slightly interesting. The slightly interesting problems are based on the following.

Theorem. First, let $\Delta x = dx$. If $\Delta x = dx$ is a small number, then dy is approximately equal to Δy.

We reach this conclusion as follows. Since

$$\lim_{\Delta x \to 0} \frac{f(x + \Delta x) - f(x)}{\Delta x} = f'(x).$$

It follows that for small values of Δx, we have

$$\frac{f(x + \Delta x) - f(x)}{\Delta x} \simeq f'(x).$$

This can be rewritten as

$$f(x + \Delta x) - f(x) \approx f'(x)\Delta x.$$

If $\Delta x = dx$, then

$$\Delta y = f(x + \Delta x) - f(x) \approx f'(x)dx = dy.$$

Clearly, the smaller Δx, the better the approximation $\Delta y \approx dy$.

Example 3. Let $y = f(x) = x^3$, find Δy and dy. Compare the values of Δy and dy when $\Delta x = dx = 2/10$ and $x = 2$.

Solution. First, $dy = 3x^2 dx$. Second, $\Delta y = (x + \Delta x)^3 - x^3 = 3x^2(\Delta x) + 3x(\Delta x)^2 + (\Delta x)^3$. Note that $\Delta y - dy = (\Delta x)^2(3x + \Delta x)$. When $x = 2$ and $dx = 2/10$, then

$$dy = 3(2)^2(2/10) = 2.4.$$

When $x = 2$ and $\Delta x = 2/10$, then

$$\Delta y = 3(2)^2(2/10) + 3(2)(2/10)^2 + (2/10)^3$$
$$= 2.648.$$

$$\Delta y - dy = (2/10)^2[6 + (2/10)] = (0.04)(6.2) = 0.248.$$

Example 4. Let $y = f(x) = \sqrt{5 + x^2}$, find Δy and dy. Compare the values of Δy and dy when both $\Delta x = -2/10$ and $dx = -2/10$ and $x = 2$.

Solution. First, $dy = x(5 + x^2)^{-1/2}dx$. When $x = 2$ and $dx = -2/10$,

$$dy = 2(5 + 4)^{-1/2}(-2/10) = -4/30 = -0.133.$$

Second, $\Delta y = \sqrt{5 + (x + \Delta x)^2} - \sqrt{5 + x^2}$. We can not really simplify this expression for Δy. Substituting $x = 2$ and $\Delta x = -2/10$, we get

$$\Delta y = \sqrt{5 + (1.8)^2} - 3 = -0.129.$$

Note that Δy and dy are approximately equal, that is, $\Delta y - dy = 0.004$.

Example 5. Let $y = f(x) = 6x^{2/3} + 3x^{4/3}$. Find the differential dy and the increment Δy in the case $\Delta x = dx = -0.8$ and $x = 64$.

Solution. First, let us find the increment Δy.

$$\Delta y = 6(x + \Delta x)^{2/3} + 3(x + \Delta x)^{4/3} - 6x^{2/3} - 3x^{4/3}.$$

This does not simplify. When $x = 64$ and $\Delta x = -0.8$ this expression for Δy becomes

$$\Delta y = 6(63.2)^{2/3} + 3(63.2)^{4/3} - 6(64)^{2/3} - 3(64)^{4/3}.$$

After some calculations we find that

$$\Delta y = -13.5749.$$

Next, let us find the differential dy.

$$dy = (4x^{-1/3} + 4x^{1/3})dx.$$

Substituting $x = 64$ and $dx = -0.8$ into this expression for dy, we get

$$dy = [4(64)^{-1/3} + 4(64)^{1/3}](-0.8)$$
$$= -13.6.$$

We are pretending that using dy to approximate Δy is an interesting problem and that replacing Δy with dy does not introduce large errors. Actually the real purpose of these exercises is just to make you compute dy and Δy. In the future when you are asked to find the differential dy or the increment Δy you will know what to do.

Example 6. A large spherical ball of radius 25 feet is to be painted with a layer of waterproofing paint 1/200 feet thick (about 1/16 inch). Find the volume of paint required.

Solution. First, the volume of a sphere is

$$V(r) = (4/3)\pi r^3.$$

The radius of the unpainted sphere is 25 feet and the radius of the painted sphere is $25 + (1/200) = 25.005$ feet. The volume of the unpainted sphere is $V(25) = (4/3)\pi(25)^3$. The volume of the painted sphere is $V(25.005) = (4/3)\pi(25.005)^3$. The volume of paint is

$$V(25.005) - V(25).$$

This is the increment ΔV when $r = 25$ and $\Delta r = 1/200$.

$$\Delta V = V(25.005) - V(25).$$

Let us approximate ΔV with dV.

$$dV = 4\pi r^2 dr.$$

Substituting $r = 25$ and $dr = 0.005$, we get

$$dV = 4\pi(25)^2(0.005) = 39.27.$$

We use this value to approximate ΔV, that is, $\Delta V \simeq 39.27$. It takes 39.27 ft^3 of paint to paint the ball. We are saying that a change in radius of dr produces a corresponding change in volume of dV.

Example 7. Suppose the position s of a particle on a line is given by $s = 2t^3 - 5t^2 + 8t + 3$, where s is in feet and t is in minutes. How far does the particle move during the time interval from $t = 2$ to $t = 2.05$ minutes? Use ds to find a change in s corresponding to a change of dt in t.

Solution. We need to find the increment corresponding to $t = 2$ and $\Delta t = 0.05$, that is

$$\Delta s = s(2.05) - s(2).$$

Let us approximate the increment Δs by the differential ds.

$$ds = (6t^2 - 10t + 8)dt.$$

When $t = 2$ and $dt = 0.05$, we have

$$ds = [6(4) - 10(2) + 8](0.05) = (12)(0.05) = (0.6).$$

During this time interval the particle moves approximately 0.6 feet.

Let us review some of the things we have just learned. First, we learned to find the differential of a function. The differential of $f(x)$ is $f'(x)dx$. The differential $f'(x)dx$ is a function of two variables x and dx. It is important to know the meaning of differential and to be able to find a differential. If we can find a derivative, it is very easy to find the differential. The fact that a differential looks very much like a derivative is a problem. If we do not have to find many differentials, we forget and think that the differential is the same as the derivative. The first purpose of this discussion is to make everyone find a few differentials. Given the function $\sin x$ the derivative is $\cos x$, but the differential is $(\cos x)dx$.

Second, we learned to find the increment of a function. Given the function $f(x)$ the increment is $\Delta f = f(x + \Delta x) - f(x)$. The increment is clearly a function of two variables x and Δx.

Third, we discussed the differential and the increment in the same discussion. In this discussion, we always assumed that $dx = \Delta x$. When this is the case the increment Δf is approximately equal to the differential df for a given value of x, that is, $\Delta f \approx df$. The point of many examples and exercises is that $\Delta f \approx df$. One reason we try to make this point is that in many situations people replace Δf with df. In fact, it is common practice in many physics texts to write dx and df when they mean Δx and Δf. This is the way Newton did calculus.

However, using dy to approximate Δy should not be viewed as an approximation method. Later we will discuss linear approximations of functions. Since the arithmetic of linear approximations is very similar to the arithmetic of differentials we sometimes confuse finding differentials and finding linear approximations. These should be viewed as different because the reason for finding differentials is not the same as the reason for finding linear approximations.

Exercises

1. Let $y = f(x) = \sqrt{16 + x^2}$. Find the differential dy when $x = 3$ and $dx = 0.2$.

2. Let $y = 4x^2 + 5x$. Find the increment Δy and the differential dy. Find the values of Δy and dy when $x = 1.5$, $dx = 0.2$, and $\Delta x = 0.2$. Find $|\Delta y - dy|$ when $x = 1.5$ and $dx = \Delta x = 0.2$.

3. Let $y = f(x) = (1/4)x^2 + 3/x$. Find the increment Δy and the differential dy. Find the values of Δy and dy when $x = 3$, $dx = 0.2$ and $\Delta x = 0.2$.

4. Use the fact that dy is approximately equal to Δy to find an approximate value of $\sqrt[3]{27.8}$. Use $\Delta y = \sqrt[3]{27.8} - \sqrt[3]{27}$.

5. Let $f(x) = 4(x+9)^{1/2} + 2x^{3/2}$. Find the increment Δy and the differential dy in the case $x = 16$ and $\Delta x = dx = -0.6$.

6. The hypotenuse of a right triangle is known to be exactly 25 inches. One of the angles is measured to be $30°$ with a maximum error of $1°$. Use differentials to estimate the maximum errors in the sides opposite and adjacent to the measured angle.

1 degree $= 0.01745$ radians, opposite error $= 0.3779$, adjacent error $= 0.2182$.

7. The outside diameter of a spherical container is 12 feet. If the wall of the container is 1/4 inch thick, use differentials to approximate the volume of the wall. What is the approximate volume of the region interior to the container?

8. Suppose the position s of a particle on a line is given by $s = t^3 - 4t^2 + 11t + 15$, where s is measured in feet and t is in minutes. How far does the particle move during the time interval from $t = 3$ to $t = 3.1$ minutes?

9. A spherical balloon is being inflated with gas. Use differentials to approximate the increase in surface area of the balloon if the radius changes from 3 ft to 3.04 ft.

6351 Related Rates

Suppose we are given a function $x(t)$ where t is time measured in seconds or minutes. The derivative with respect to t, $x'(t)$, is the rate of change of x with respect to t, time. We have already considered one such situation. If $x(t)$ denotes the position of a particle moving back and forth along a straight line, then $x'(t)$ denotes the velocity of the particle. Velocity is the rate of change of position with respect to time. If position is measured in feet and time is measured in seconds, then $x'(t)$, the velocity, is measured in feet per second. This is an example of a rate problem. We are now going to consider some other problems involving rates. In these problems we will have two or more rates connected by an equation. These problems are usually called related rates problems. This is because the problems involve two or more functions that depend on time and then are connected by an equation. This means that the rate at which one function changes is actually determined by (is related to) the rate at which the other function changes.

Example 1. Suppose air is being pumped into a balloon which always takes the form of a sphere. Suppose air is being pumped into the balloon at the rate of 60 cm^3/sec. How fast is the radius of the balloon increasing when the radius is 40 cm?

Solution. Let r denote the radius of the balloon and V denote the volume. The radius and volume of a sphere are related by the equation

$$V = (4/3)\pi r^3.$$

We are not dealing with a static sphere. Both the radius and the volume are changing with time. Let us indicate this by writing the volume as $V(t)$ and the radius as $r(t)$. Therefore,

$$V(t) = (4/3)\pi[r(t)]^3.$$

The rate at which the radius is changing is $\dfrac{dr}{dt}$ and the rate at which the volume is changing is $\dfrac{dV}{dt}$. Let us differentiate both sides of the equation with respect to t. We will need to use the chain rule.

$$\frac{dV}{dt} = (4/3)\pi \frac{d}{dt}[r(t)]^3$$

$$\frac{dV}{dt} = (4/3)\pi \frac{d}{dr}[r^3]\frac{dr}{dt}$$

$$\frac{dV}{dt} = (4/3)\pi(3)r^2\frac{dr}{dt}$$

$$\frac{dV}{dt} = 4\pi[r(t)]^2\frac{dr}{dt}$$

This is an equation which relates the rate of change of volume to the rate of change of radius.

We are not given a value of t. This is good since we do not know either the function $r(t)$ or the function $V(t)$. Instead, we are told that at whatever time it is, say $t = t_0$, then $r(t_0) = 40$ and $\frac{dV}{dt}\big|_{t=t_0} = 60$. We must find $\frac{dr}{dt}\big|_{t=t_0}$, the rate of change of radius with respect to time at this instant in time. If we let $t = t_0$ in the above equation, we get

$$V'(t_0) = 4\pi[r(t_0)]^2 r'(t_0).$$

Substituting in the numerical values,

$$60 = 4\pi[40]^2 r'(t_0)$$

$$r'(t_0) = \frac{dr}{dt}\bigg|_{t=t_0} = \frac{60}{4\pi(1600)} = \frac{3}{320\pi} \text{ cm/sec.}$$

All related rates problems follow this general pattern. First, we find that we have two or more variables related by an equation. For example, the equation relating two variables might be

$$x^3 + y^2 = 100.$$

The values of these variables x and y are functions of time so that we write

$$[x(t)]^3 + [y(t)]^2 = 100.$$

We next differentiate both sides of this equation with respect to t. In order to find $\frac{d}{dt}[x(t)]^3$ we must note that $[x(t)]^3$ is a composite function. The inside function is $x(t)$. We use the chain rule $\frac{d}{dt}(x^3) = \frac{d}{dx}[x^3]\frac{dx}{dt}$.

$$3[x(t)]^2 x'(t) + 2[y(t)]y'(t) = 0.$$

Note, that this resulting equation involves rates which are given by derivatives. In order to do this differentiation we almost always have to use the chain rule as we did here. Differentiation gives us an equation involving the original variables $x(t)$ and $y(t)$ and their respective rates of change $x'(t)$ and $y'(t)$. The statement of the problem then gives us all these quantities except one at some instant of time t and asks us to solve for the one not given.

Example 2. A very long ladder 125 feet long rests against a vertical wall. Suppose the bottom end of the ladder slides along the floor away from the wall at the rate of 15 ft/sec. How fast is the top of the ladder sliding down the wall when the bottom of the ladder is 75 feet from the wall? We assume that the top of the ladder remains in contact with the wall.

Solution. We always must start by naming the unknown quantities. We must select a letter to denote any quantity which is changing with time during the action described in the problem. The length of the ladder does not change. Let x denote the distance from the wall to the base of the ladder. Let y denote the distance from the floor to the top of the ladder. We need an equation which relates the quantities that change with time. A simple diagram is often helpful.

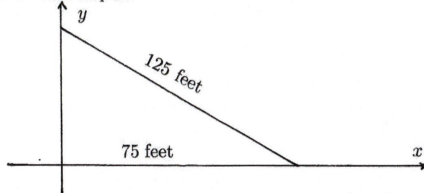

This is a right triangle. The Pythagorean theorem says $x^2 + y^2 = 125^2$. It is clear that both x and y change with time. To remind us of this fact, we rewrite the equation as

$$[x(t)]^2 + [y(t)]^2 = 125^2.$$

Note that this equation relates functions which are not known. This is the case in most related rates problems. This means that we will need the chain rule to differentiate.

Differentiate both sides of this equation with respect to t.

$$\frac{d}{dt}[x(t)]^2 + \frac{d}{dt}[y(t)]^2 = \frac{d}{dt}[125^2].$$

Use the chain rule to differentiate and we get

$$2x(t)x'(t) + 2y(t)y'(t) = 0.$$

We are given that at some instant in time, say $t = t_0$, then $x(t_0) = 75$ and $x'(t_0) = 15$. Since we know $x(t_0)$, we can find $y(t_0)$ from the equation

$$[x(t_0)]^2 + [y(t_0)]^2 = 125^2$$
$$75^2 + [y(t_0)]^2 = 125^2$$
$$y(t_0) = 100.$$

Substituting into $x(t_0)x'(t_0) + y(t_0)y'(t_0) = 0$, we get

$$(75)(15) + (100)y'(t_0) = 0$$

$$y'(t_0) = -11.25.$$

The variable y represents the distance from the floor to the top of the ladder. The fact that the rate $y'(t_0)$ is negative indicates that the velocity is in the downward direction. The top of the ladder is sliding down at the rate of 11.25 ft/sec at this instant of time.

We can also solve this problem directly by assuming a certain time as the initial time and a certain position as the initial position. However, no initial time and no initial position are given.

Example 3. Two long straight highways intersect at right angles. One highway goes east and west while the other highway goes north and south. At a certain time car A is 45 miles west of the intersection traveling east at 65 mph. Two hours later car B starts at the intersection and travels north at 70 mph. At what rate is the distance between the two cars changing 3 hours after car B leaves the intersection.

Solution. Let $x(t)$ denote the distance that car A is east of the intersection at time t and let $y(t)$ denote the distance that car B is north of the intersection at time t. Let $z(t)$ denote the distance between the two cars at time t. We need a letter to represent each of the quantities that change with time. A simple diagram:

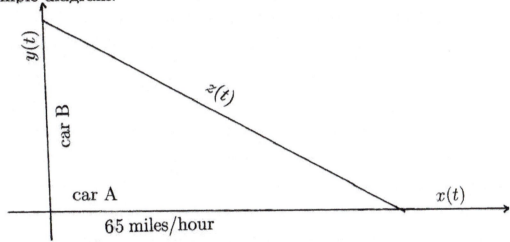

This is a right triangle. The Pythagorean theorem says

$$[x(t)]^2 + [y(t)]^2 = [z(t)]^2.$$

Note that this equation is correct even when $x(t) < 0$ or $y(t) < 0$. However, if for example $x(t) < 0$, then $x(t)$ is not an actual length or distance. If we use $x(t) < 0$ at the time of the action, then a more careful analysis is required. Differentiate both sides of this equation using the chain rule. We get

$$2x(t)x'(t) + 2y(t)y'(t) = 2z(t)z'(t).$$

Let $t = t_0$ denote the instant in time in question, then

$$x(t_0)x'(t_0) + y(t_0)y'(t_0) = z(t_0)z'(t_0).$$

If $t = 0$ is the time when car B leaves the intersection going north, then $t_0 = 3$. If $t = 0$ is the time when car A is 45 miles west of the intersection, then $t_0 = 5$. Clearly, $x(t_0) = 5(65) - 45 = 280$ and $y(t_0) = 3(70) = 210$. We can compute $z(t_0)$ from $x^2 + y^2 = z^2$.

$$(280)^2 + (210)^2 = [z(t_0)]^2$$
$$z(t_0) = 350.$$

Substituting for $x(t_0)$, $y(t_0)$, $z(t_0)$, $x'(t_0)$, and $y(t_0)$, we get

$$(280)(70) + (210)(65) = (350)[z'(t_0)]$$
$$95 = z'(t_0).$$

The distance between the cars is increasing at the rate of 95 miles/hour at the instant of time in the question.

Example 4. A water tank has the shape of an inverted circular cone (an ice cream cone). The radius of the top of the tank is 10 feet and the height is 35 feet. If water is being pumped into the tank at the rate of 20 ft^3/min, find the rate at which the water level is rising when the water is 21 feet deep.

Solution. What measurements are changing with time? First, there is the volume of water in the tank at time t. The body of water in the tank is always in the shape of a cone. Both the height and radius of this cone of water changes with time. Let $h(t)$ denote the height (depth) of the cone of water in the tank. Let $r(t)$ denote the radius of the top of the cone of water in the tank. Let $V(t)$ denote the volume of water in the tank. We need a relationship between $V(t)$, $h(t)$, and $r(t)$. Let us draw a cross section of the cone.

The volume of the cone of water is given by $V = (1/3)\pi r^2 h$.

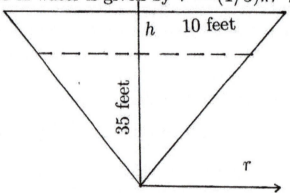

Draw cartesian coordinate lines with the tip of the triangle (cone) as the origin. Draw an h axis which is vertical. Draw an r axis parallel to the ground through the tip of the cone. In order to answer the question we need an equation for V in terms of h. This means we need a relationship connecting r, the radius of the cone of water, and h, the depth of the cone of water. The edge of the triangle is part of a line through $(0,0)$ and $(10, 35)$. The slope of this line is $35/10$. The equation of the line is

$$h = \frac{7}{2}r.$$

Alternately we can use similar triangles to get

$$\frac{r}{10} = \frac{h}{35}$$
$$r = (2h/7)$$

This equation enables us to write the volume of water in the cone at any time t as

$$V = (1/3)\pi(2h/7)^2 h$$

Since V and h both depend on time we write this equation as

$$V(t) = (4/147)\pi[h(t)]^3.$$

Now that we have only $h(t)$ in the equation we differentiate both sides of this equation with respect to t, time. We use the chain rule to differentiate $[h(t)]^3$.

$$V'(t) = (4/147)\pi(3)[h(t)]^2 h'(t)$$
$$V'(t) = (4/49)\pi[h(t)]^2 h'(t).$$

Let t_0 denote the instant in time when the water is 21 feet deep, that is, $h(t_0) = 21..$ We are given that $V'(t_0) = 20$ ft^3/min. Substituting these values into the above equation, we get

$$20 = (4/49)\pi[21]^2 h'(t_0)$$
$$h'(t_0) = \frac{20}{36\pi} = \frac{5}{9\pi} \text{ ft/min.}$$

The height of water in the tank is increasing at the rate of $5/9\pi \approx 0.1768$ ft/min.

Example 5. A searchlight is rotating clockwise about a point which is 50 feet from a long straight wall. As the light rotates the spot of light cast by the searchlight moves along the wall. Mark the spot on the wall which is closest to the searchlight. A line from this closest point to the searchlight makes a right angle with the wall. How fast is the spot moving along the wall when the spot is 20 feet from the marked closest point? Suppose the searchlight is rotating at 3 revolutions per minute.

Solution. Let us draw a diagram.

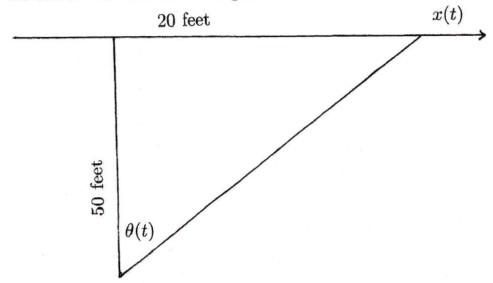

What expressions in this problem depend on time? Let $x(t)$ denote the distance from the closest point on the wall to the point on the wall where the light beam strikes the wall. Let $\theta(t)$ denote the angle of rotation of the searchlight where $\theta = 0$ indicates the angle when the searchlight is shining directly on the marked closest point. Using this notation, $\frac{dx}{dt}$ is the rate at which the lighted spot moves down the wall and $\frac{d\theta}{dt}$ is the rate at which the angle θ changes. We are given that the searchlight rotates at the rate of 3 revolutions per minute. But θ is measured in radians. This means

$$\frac{d\theta}{dt} = 3(2\pi) = 6\pi \text{ radians/min.}$$

From the diagram we see that

$$\tan \theta = \frac{x}{50} \text{ or } x = 50 \tan \theta,$$
$$x(t) = 50 \tan[\theta(t)].$$

We differentiate with respect to t using the chain rule.

$$x'(t) = 50 \sec^2[\theta(t)]\theta'(t).$$

At the instant in time in question

$$\sec\theta = \frac{\text{hypotenuse}}{\text{adjacent}} = \frac{\sqrt{50^2 + 20^2}}{50} = \frac{\sqrt{29}}{5}.$$

Substituting we get

$$x'(t) = 50[\sqrt{29}/5]^2(6\pi)$$
$$= 2(29)(6\pi) = 348\pi \approx 1093 \text{ ft/min.}$$

Note that in most related rates problems we need to differentiate something like $\tan[\theta(t)]$ with respect to t. Since the function $\theta(t)$ is unknown we always need to use the chain rule to perform this differentiation. Related rate problems give us a chance to practice the chain rule.

Exercises

1. A very long ladder 150 feet long, rests against a vertical wall. Suppose the ladder slides away from the wall at the rate of 20 ft/sec. How fast is the top of the ladder sliding down the wall when the bottom of the ladder is 90 feet from the wall? Assume that the top of the ladder remains in contact with the wall.

2. A stone dropped into a still pond sends out a circular ripple whose radius increases at a constant rate of 4 ft/sec. How rapidly is the area enclosed by the ripple increasing at the end of 8 seconds?

3. Two cars are traveling on long straight roads that meet at right angles. The first car leaves the intersection and travels north at 32 miles per hour. The second car leaves the intersection at a different time and travels east at 51 miles per hour. At a certain moment in time, the car going north is 240 miles from the intersection and the car going east is 320 miles from the intersection. At that moment, at what rate is the distance between the two cars increasing.

4. Two cars are traveling on long straight roads that meet at right angles. Car A leaves the intersection traveling east at 48 miles per hour. Car B leaves the intersection 3 hours later and travels north at 50 miles per hour. At what rate is the distance between the two cars increasing 2 hours after car B leaves the intersection?

5. A rocket, rising vertically, is tracked by a radar station that is 6 miles from the launch pad. When the rocket is 4 miles high its speed is 30 miles per minute. At what rate is the angle of elevation of the radar beam increasing at this instant in time.

6. A kite 100 ft above ground moves horizontally in a line at a speed of 10 ft/sec. At what rate is the angle between the string and the horizontal decreasing when 200 ft of string has been let out? At what rate is the distance between the person holding the kite and the kite changing when the string is 200 feet long.

7. Two cars are traveling along long straight roads which meet at right angles. Car A starts from a point 70 miles west of the intersection and travels east at 65 miles per hour. At the same moment car B starts from a point 120 miles north of the intersection and travels north at 60 miles per hour. At what rate is the distance between the two cars changing 6 hours later?

8. A water tank has the shape of a circular cone pointed end points down with the radius of the top of 12 feet and height 20 feet. If water is pumped into the tank at the rate of 10π ft^3/min, find the rate at which the water level is rising when the water is 15 feet deep.

9. A water tank has the shape of a circular cone pointed end points up. The radius of the bottom is 12 feet and the height is 20 feet. If water is pumped into the tank at the rate of 10π ft^3/min, find the rate at which the water level is rising when the water is 15 feet deep.

$$\frac{dh}{dt} = 10/9 \text{ ft/sec}$$

10. Sand is being dumped from a conveyor belt at the rate of 18π ft^3/min. The courseness of the sand is such that it forms a pile in the shape of a cone

with the radius of the base always 1/3 the height. How fast is the height increasing when the pile is 15 feet high?

11. A trough is 10 feet long and a cross section has the shape of an isosceles trapezoid that is 24 feet wide at the top and 14 feet wide at the bottom. The trough is 12 feet deep. If the trough is filled with water at the rate of 120 ft^3, how fast is the water level rising when the water is 8 feet deep?

12. A lighthouse is located 400 yards away from the nearest point P on a straight shoreline. The light makes three revolutions per minute. How fast is the beam of light moving along the straight shoreline when the point where the beam strikes the shore is 80 yards from point P.

13. A man six feet tall is walking at the rate of 3 ft/sec toward a street light 24 feet high. At what rate is the length of his shadow changing when the man is 15 feet from the base of the light pole?

14. A plane flies directly over an observer on the ground at 600 feet per second. The plane flies exactly at 2500 feet high above level ground. Exactly 10 seconds after the plane passes over the observer at what rate is the angle of elevation of the plane with respect to the observer changing?

15. A particle is moving along the curve $y = \sqrt{9 + x^2}$. At the point $(4,5)$ the x coordinate is increasing at the rate of 3 cm/sec. How fast is the y coordinate of the point changing at this instant in time?

$$\frac{dy}{dt} = (12/5)cm/sec$$

16. A long ladder 250 feet long is leaning against a wall with one end on the floor. The bottom end of the ladder is being pulled along the floor. At a certain instant of time the bottom end of the ladder is 90 feet from the wall, the velocity of the bottom end is 32 ft/sec, and the acceleration is 18 ft/sec^2 away from the wall. Assume that the top of the ladder is always in contact with the wall. What are the velocity and acceleration of the top of the ladder at this instant in time?

6353 Increasing and Decreasing for Functions

We are going to discuss some properties of functions. But first let us think a bit about the values of x for which a given function is defined. Our first thought when we say "given a function $f(x)$" is that the function is defined for all values of x. Below is a list of functions that are indeed defined for every real number x.

$$3x^4 + 5x^3 + 7x$$

$$e^{3x} + x^2$$

$$\sin 2x + \cos 5x.$$

On the other hand there are some functions which are not defined for every real number. We just have in mind functions for which there are a few real numbers for which the function is not defined. A few such functions are listed:

$\dfrac{1}{x-5}$ is not defined for $x = 5$.

$\dfrac{x^2 + 2}{(x+2)(x-4)}$ is not defined for $x = -2$ and $x = 4$.

$\ln|x|$ is not defined for $x = 0$.

$\tan x$ is not defined for $x = \pi/2,\ 3\pi/2, 5\pi/2$, etc.

Some functions are defined for an interval of values of x, but not defined for other intervals of values of x. Some examples of such functions are:

$\sqrt{x} = x^{1/2}$ is defined for $x \geq 0$ but not defined for $x < 0$.

$\ln x$ is defined for $x > 0$ but not defined for $x \leq 0$.

$\sqrt{25 - x^2}$ is defined for $-5 \leq x \leq 5$, but not defined for $x < -5$ and $x > 5$.

If the values of x for which a given function is defined are a critical issue, then we will discuss it. Otherwise, our approach will be that when a particular function $f(x)$ is mentioned it is expected that the reader will determine for what values of x the function is defined using his past experience. All the values of x taken together for which a function is defined is often called the "domain" of the function.

When we wish to discuss a general function $f(x)$ without specifying exactly which particular function we mean, then we say something like "suppose $f(x)$ is a given function". In this case $f(x)$ can represent any of our well known functions. If the function is to have some particular property, then we state that property. For example we might say "suppose $f(x)$ is defined for $c \le x \le d$". If we wish to consider a general value of x, we say something like "suppose $x = b$". When we say "suppose $x = b$" we understand that the function is defined when $x = b$. In almost all cases when we say that a function is defined at $x = b$ we also want the function defined for values of x such that $x < b$ and such that $x > b$. These values of x may need to be very close to b, that is, the function is defined for all values of x such that $|x - b|$ is a small number.

We often like to compare values of a function $f(x)$. This is especially true if we are dealing with a practical problem so that the values of $f(x)$ represent something concrete. If x represents the number of units of a product produced and $f(x)$ represents the profit from x units, then we would certainly be interested in the question: As the value of x increases does the value of $f(x)$, profit, increase? In some cases we may just want to compare values.

We are now going to consider the questions of for what values of x is a function increasing.

Example 1. Suppose $f(x) = x^3 + 2x^2 - 3x$, is it true that $f(1)$ is larger than $f(3)$?

Solution. Substituting we get $f(1) = 0$ and $f(3) = 39$. Clearly, $f(1) < f(3)$. The value of $f(x)$ increased as x moved from 1 to 3.

Definition of increasing at a point. Given $f(x)$ and a number c if there exists a number $d > 0$ such that $f(x)$ is defined for all x such that $|x - c| < d$ and if $f(x)$ has the following properties.

(a) For any number x_1 such that $c - d < x_1 < c$, we have $f(x_1) < f(c)$.

(b) For any number x_2 such that $c < x_2 < c + d$, we have $f(c) < f(x_2)$.

Then we say that the function $f(x)$ is increasing at the point where $x = c$.

This definition says that $f(x)$ is increasing at $x = c$ if $f(x) < f(c)$ for $x < c$ and $f(x) > f(c)$ for $x > c$, that is, the values of $f(x)$ are getting larger and larger as x increases when x is near c.

Note that in order to say that a function is increasing at a point $x = c$ the function must be defined for all x such that $|x - c| < d$ for some small positive number d. When this is the case we say "the number c is an interior point of the domain of the function $f(x)$".

In general when discussing the question of whether a function is increasing or decreasing we will always assume that the function is continuous.

Example 2. Let $f(x) = x^3 + 2x^2$ and let c denote any number such that $c > 0$. Show that $f(x)$ is increasing at $x = c$.

Solution. Let c denote a number such that $c > 0$. First, let x_1 denote a value of x such that $0 < x_1 < c$. Note that if c is close to zero, then x_1 is closer to zero. We have

$$
\begin{aligned}
f(c) - f(x_1) &= [c^3 + 2c^2] - [x_1^3 + 2x_1^2] \\
&= [c^3 - x_1^3] + [2c^2 - 2x_1^2] \\
&= [(c - x_1)(c^2 + cx_1 + x_1^2)] + [2(c - x_1)(c + x_1)] \\
&= (c - x_1)[c^2 + cx_1 + x_1^2 + 2c + 2x_1].
\end{aligned}
$$

Since $c > 0$ and $x_1 > 0$ it follows that $c^2 + cx_1 + x_1^2 + 2c + 2x_1 > 0$. Also $c > x_1$. The product of two positive numbers is positive.

$$
\begin{aligned}
f(c) - f(x_1) &> 0 \\
f(c) &> f(x_1) \\
f(x_1) &< (f(c).
\end{aligned}
$$

Second, let x_2 denote a value of x such that $x_2 > c$.

$$
\begin{aligned}
f(x_2) - f(c) &= [x_2^3 + 2x_2^2] - [c^3 + 2c^2] \\
&= [x_2^3 - c^3] + 2[x_2^2 - c^2] \\
&= [(x_2 - c)(x_2^2 + x_2c + c^2)] + 2[(x_2 - c)(x_2 + c)] \\
&= (x_2 - c)[x_2^2 + x_2c + c^2 + 2x_2 + 2c]
\end{aligned}
$$

128

Since $x_2 > c > 0$, it follows that $x_2^2 + x_2 c + c^2 + 2x_2 + 2c > 0$. Also $x_2 - c > 0$. The product of two positive numbers is positive.

$$f(x_2) - f(c) > 0$$
$$f(x_2) > f(c).$$

Note that the number d in the definition of increasing at a point can be any positive number. In this case we have $f(x_2) > f(c)$ no matter what positive number we choose for d. The number x_2 need not be close to c.

According to the definition of increasing at a point this means that $f(x)$ is an increasing function at $x = c$ for any $c > 0$.

Definition of decreasing at a point. Given $f(x)$ and a number c, if there exists a number $d > 0$ such that $f(x)$ is defined for all x such that $|x - c| < d$ and if $f(x)$ has the following properties.

(a) For any number x_1 such that $c - d < x_1 < c$, we have $f(x_1) > f(c)$.

(b) For any number x_2 such that $c < x_2 < c + d$, we have $f(c) > f(x_2)$.

Then we say that the function $f(x)$ is decreasing at the point where $x = c$.

This definition says that $f(x)$ is decreasing at $x = c$ if $f(x) > f(c)$ for $x < c$ and $f(c) > f(x)$ for $x > c$, that is, the values of $f(x)$ are getting smaller and smaller as x increases when x is near c.

Example 3. Let $f(x) = x^2 - 4x$ and c any number such that $c < 2$. Show that $f(x)$ is decreasing at $x = c$.

Solution. Let c be given such that $c < 2$. First, let x_1 denote a number such that $x_1 < c$. Note that $x_1 < 2$ since $c < 2$. We have

$$\begin{aligned}
f(x_1) - f(c) &= [x_1^2 - 4x_1] - [c^2 - 4c] \\
&= [x_1^2 - c^2] + [-4x_1 + 4c] \\
&= [(x_1 - c)(x_1 + c)] - 4(x_1 - c) \\
&= (x_1 - c)[x_1 + c - 4].
\end{aligned}$$

Since $x_1 < 2$ and $c < 2$, it follows that $x_1 + c < 4$ or $x_1 + c - 4 < 0$. Also $x_1 - c < 0$. The product of two negative numbers is a positive number. Thus

$$f(x_1) - f(c) > 0$$

$$f(x_1) > f(c) \text{ for } x_1 < c.$$

Second, let x_2 denote a number such that $x_2 > c$ and $x_2 < 2$ or $c < x_2 < 2$. If c is close to 2, then x_2 is closer to 2.

$$\begin{aligned}
f(x_2) - f(c) &= [x_2^2 - 4x_2] - [c^2 - 4c] \\
&= [(x_2 - c)(x_2 + c)] - 4(x_2 - c) \\
&= (x_2 - c)[x_2 + c - 4]
\end{aligned}$$

Since $x_2 < 2$ and $c < 2$, it follows that $x_2 + c - 4 < 0$. Also $x_2 - c > 0$. The product of a negative number and a positive number is a negative number. Thus

$$f(x_2) - f(c) < 0$$
$$f(x_2) < f(c) \text{ when } c < x_2 < 2.$$

According to the definition of decreasing at a point this means that $f(x)$ is a decreasing function at $x = c$ for $c < 2$. Note that in this case we must choose d such that $0 < d < 2 - c$ since $|x_2 - c| < 2 - c$. We must choose x_2 close to c.

Theorem 1. If $f'(c) > 0$, then $f(x)$ is increasing at the point $x = c$.

Proof. Suppose c is such that $f'(c) > 0$. Recall that when we say that $f'(c)$ exists, then it must be true that $f(x)$ is defined for all x such that $|x - c|$ is sufficiently small. The definition of the derivative of $f(x)$ at $x = c$ is

$$f'(c) = \lim_{x \to c} \frac{f(x) - f(c)}{x - c}.$$

We are supposing that $f'(c) > 0$. This says that

$$\lim_{x \to c} \frac{f(x) - f(c)}{x - c} > 0.$$

Since the limit is a positive number this means that there exists a small positive number d such that for values of x such that $|x - c| < d$, then

$$\frac{f(x) - f(c)}{x - c} > 0.$$

Let x_1 denote any value of x such that $c - d < x_1 < c$. Substituting x_1 for x we have

$$\frac{f(x_1) - f(c)}{x_1 - c} > 0.$$

Since $x_1 - c < 0$, this says

$$\frac{f(x_1) - f(c)}{\text{negative}} = \text{positive or } f(x_1) - f(c) = \text{negative.}$$

It must be true that $f(x_1) - f(c) < 0$ or $f(x_1) < f(c)$.

Let x_2 denote any value of x such that $c < x_2 < c + d$. Substituting x_2 for x we have

$$\frac{f(x_2) - f(c)}{x_2 - c} > 0.$$

Since $x_2 - c > 0$, this says

$$\frac{f(x_2) - f(c)}{\text{positive}} = \text{positive or } f(x_2) - f(c) = \text{positive.}$$

It must be true that $f(x_2) - f(c) > 0$ or $f(x_2) > f(c)$.

Thus the two conditions in the definition of increasing are satisfied by $f(x)$ at $x = c$, and so we say that $f(x)$ is increasing at $x = c$.

Theorem 2. If $f'(c) < 0$, then $f(x)$ is decreasing at $x = c$.

Proof. Suppose c is such that $f'(c) < 0$. Recall the definition of the derivative of $f(x)$ at $x = c$.

$$f'(c) = \lim_{x \to c} \frac{f(x) - f(c)}{x - c}.$$

We are supposing that $f'(c) < 0$. This says that

$$\lim_{x \to c} \frac{f(x) - f(c)}{x - c} < 0.$$

Since the limit is a negative number there exists a small number d such that $\frac{f(x) - f(c)}{x - c} < 0$ for all x such that $|x - c| < d$ or $c - d < x < c + d$, we have

$$\frac{f(x) - f(c)}{x - c} < 0.$$

Let x_1 denote any value of x such that $c - d < x_1 < c$, then $\frac{f(x_1) - f(c)}{x_1 - c} < 0$. Since $x_1 - c < 0$, this says that

$$\frac{f(x_1) - f(c)}{\text{negative}} = \text{negative}.$$

It must be true that $f(x_1) - f(c) > 0$ or $f(x_1) > f(c)$.

Let x_2 denote any value of x such that $c < x_2 < c + d$, then $\frac{f(x_2) - f(c)}{x_2 - c} < 0$. Since $x_2 - c > 0$ this says that

$$\frac{f(x_2) - f(c)}{\text{positive}} = \text{negative}.$$

It must be true that $f(x_2) - f(c) < 0$ or $f(x_2) < f(c)$.

Thus the conditions of the definition of decreasing are satisfied and so we say that $f(x)$ is decreasing at $x = c$.

Example 4. For what values of x is the function $f(x) = x^3 - 3x^2 - 24x$ increasing?

Solution. Using the definition of increasing is a lot of work. Instead to answer this question we will apply Theorem 1. In order to apply Theorem 1 we need to look at the derivative of $f(x)$,

$$f'(x) = 3x^2 - 6x - 24 = 3(x - 4)(x + 2).$$

For what values of x is $3(x - 4)(x + 2) > 0$? Clearly $3(x - 4)(x + 2) = 0$ when $x = 4$ and $x = -2$. Start by substituting a value of x less than -2, say $x = -3$. When $x = -3$, we get $f'(-3) = 21$, a positive number. Next substitute a value of x between -2 and 4, say $x = 0$. Let $x = 0$, then $f'(0) = -24$, a negative number. Substitute a value larger than 4. Let $x = 5$, then $f'(5) = 21$, a positive number.

The function $(x - 4)(x + 2)$ is equal to zero only when $x = -2$ and $x = 4$ and is continuous. This says the expression $f'(x) = 3(x - 4)(x + 2)$ can only change signs at $x = -2$ and at $x = 4$. Therefore, $f'(x) > 0$ for $x < -2$ and also $f'(x) > 0$ for $x > 4$. Also $f'(x) < 0$ for $-2 < x < 4$.

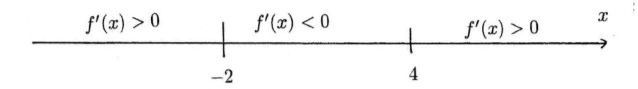

$$f'(x) > 0 \qquad f'(x) < 0 \qquad f'(x) > 0 \qquad x$$

$$-2 \qquad\qquad 4$$

We conclude that $f'(x) > 0$ either whenever $x < -2$ or whenever $x > 4$. It follows from Theorem 1 that $f(x)$ is increasing for all values of x such that either $x < -2$ or values such that $x > 4$. We could also conclude using Theorem 2 that $f(x)$ is decreasing for all value of x such that $-2 < x < 4$.

When discussing increasing and decreasing for functions it is always very helpful to look at the graph of the function in question. A great deal about a function can be learned by looking at the graph. Consider the x and y coordinates of points on the graph (curve). If the graph is increasing, then $f(x)$ gets larger as x gets larger and the function is increasing. If the y values of the points are getting larger as the x values get larger, then we say that the graph (curve) is increasing. If the graph is increasing, then the function is increasing. If the y values of points are getting smaller as the x values get larger, then we say that the graph (curve) is decreasing. If the graph is decreasing, then $f(x)$ gets smaller as x gets larger and the function is decreasing.

Example 5. Suppose we have a function $f(x)$ and the graph of the function is given below. For what values of x is $f(x)$ increasing and for what values of x is $f(x)$ decreasing?

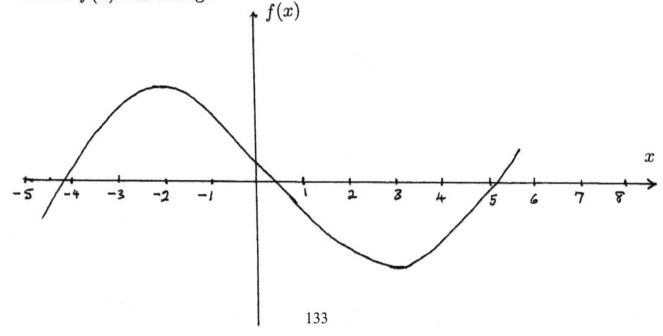

133

Solution. First, we realize that up is positive. The higher the graph along the y axis the larger the value of $f(x)$. With this in mind we are able to reach certain conclusions about for which values of x the function $f(x)$ is increasing and for which values of x the function is decreasing by just looking at the graph. If $x < -2$, the graph clearly shows that as x increases the value of $f(x)$ increases. If $x_1 < x_2 < -2$, then $f(x_1) < f(x_2)$. If $-2 < x < 3$, the graph makes it clear that as x increases the values of $f(x)$ decreases. The function $f(x)$ is decreasing for any x such that $-2 < x < 3$. If $x > 3$, as x increases the values of $f(x)$ increase. The function $f(x)$ is increasing for any x such that $x > 3$.

Example 6. Sketch the graph of a function with the following properties:

(1) If $-3 < x < 4$, then $f(x)$ is increasing.

(2) If $x < -3$ or $x > 4$, then $f(x)$ is decreasing.

Solution. We are not given any values of $f(x)$. This means that as long as we satisfy (1) and (2) we may choose our own values for $f(x)$

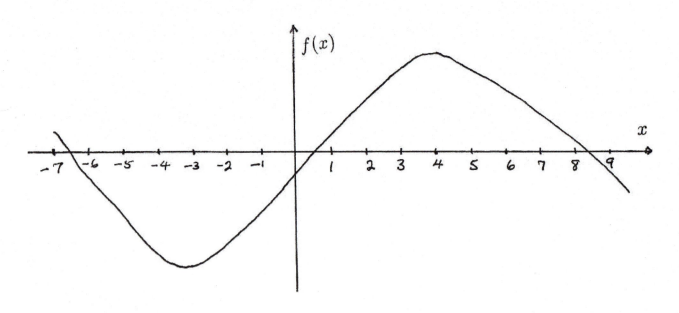

Example 7. For what values of x is the function $f(x) = (25 - x^2)^{1/2}$ increasing?

Solution. First, note that $\sqrt{25 - x^2}$ is only defined for x in the closed interval $-5 \leq x \leq 5$. This means that there is no question to answer when $x < -5$ or when $x > 5$. Next note that $f(x)$ is not defined for x on both sides of $x = -5$ and $x = 5$. Checking our definition of increasing and our definition of decreasing we see that at this time we have not defined what it means to say that $f(x)$ is increasing or $f(x)$ is decreasing for $x = -5$ or $x = 5$. We plan to use Theorem 1 to answer the question for values of x such that $-5 < x < 5$ and so we need to find the derivative.

$$f'(x) = -x(25 - x^2)^{-1/2}, \quad -5 < x < 5.$$

Note that according to this formula for the derivative $f'(-5)$ and $f'(5)$ are not defined. For $-5 < x < 5$, we have $25 - x^2 > 0$. This means that

$$f'(x) = \frac{-x}{\text{positive}}.$$

If $-5 < x < 0$, then $f'(x) > 0$. If x is any individual number in the open interval $-5 < x < 0$, then $f(x)$ is increasing. We applied Theorem 1. If $0 < x < 5$, then $f'(x) < 0$. We can apply Theorem 2 to say that if x is in the open interval $0 < x < 5$, then $f(x)$ is decreasing.

Example 8. Let $f(x) = (x - 8)^{2/3}$. For what values of x is $f(x)$ increasing and for what values of x is $f(x)$ decreasing.

Solution. Note that $f(x)$ is defined for all real numbers x. We want to apply Theorems 1 and 2. We need to find $f'(x)$,

$$f'(x) = 2/3(x - 8)^{-1/3}.$$

Note that $f'(8)$ is not defined. Let c denote a number such that $c < 8$, then $f'(c) < 0$ since $c - 8 < 0$. Theorem 2 tells us that $f(x)$ is decreasing if $x < 8$. Let c denote a number such that $c > 8$, then $f'(c) > 0$ since $c - 8 > 0$. Theorem 1 tells us that $f(x)$ is increasing for $x > 8$. Since $f'(8) = DNE$ neither theorem applies when $x = 8$. Since $f(x) > 0$ for $x \neq 8$ and $f(8) = 0$, it is clear that $f(x)$ is neither increasing nor decreasing at $x = 8$. We have $f(x) > f(8)$ for x on either side of 8.

We need to gain some understanding of functions defined in parts or sections. Consider the function

$$f(x) = \begin{cases} -2x + 1 & x < 0 \\ x^2 + 1 & 0 \le x \le 2 \\ x + 3 & x > 2. \end{cases}$$

First, let us be sure that we understand the meaning of the notation. If $x < 2$, then the value of $f(x)$ is given by $-2x+1$, that is, $f(-2) = -2(-2)+1 = 5$ or $f(-10) = 21$. If the value of x is in the interval $0 \le x \le 2$, then $f(x)$ is given by $x^2 + 1$, that is, $f(1) = 1 + 1 = 2$ or $f(1.5) = (1.5)^2 + 1 = 3.25$. If x is such that $x > 2$, then $f(x)$ is given by the expression $x + 3$, that is, $f(5) = 5 + 3 = 8$. We can sketch a graph of this function,

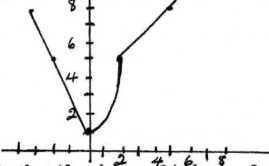

Note that if we let $x = 0$ in both $-2x+1$ and $x^2 + 1$, we get 1. The function $f(x)$ is continuous at $x = 0$. If we let $x = 2$ in both $x^2 + 1$ and $x + 3$, we get the value 5. The function $f(x)$ is continuous at $x = 2$.

Recall that the definition of derivative $f'(x)$ is a two sided limit. This means that in order for $f'(c)$ to be defined the function $f(x)$ must exist for some values of x such that $x < c$ and also some values such that $x > c$. Given $x < 0$, $f(x)$ is defined by $f(x) = -2x + 1$ for values on both sides of x. This means $f'(x) = -2$ for $x < 0$. Given x such that $0 < x < 2$ $f(x)$ is defined by $f(x) = x^2 + 1$ for values on both sides of x. This means $f'(x) = 2x$ for $0 < x < 2$. Given x such that $x > 2$ $f(x)$ is defined by $f(x) = x + 1$ for some values on both sides of x. This means that $f'(x) = 1$ for $x > 2$.

$$f'(x) = \begin{cases} -2 & x < 0 \\ 2x & 0 < x < 2 \\ 1 & x > 2. \end{cases}$$

Note that we did not find $f'(0)$ or $f'(2)$. We cannot use our regular formulas to find $f'(0)$ because $f(x)$ is given by $-2x + 1$ for $x < 0$ and an entirely different expression $x^2 + 1$ for $x > 0$. In fact, $f'(0) = DNE$. In order to show that $f'(0) = DNE$ we would need to use the definition of derivative. Also $f'(2) = DNE$.

Example 9. Let $g(x) = x|x - 6|$. For what values of x is $g(x)$ increasing and for what values of x is $g(x)$ decreasing?

Solution. Note that $g(x)$ is defined for all real numbers x. We want to apply Theorems 1 and 2. We need to find $g'(x)$. We have a difficulty in that we have never learned the formula for the derivative of $|x - 6|$. The easiest way to handle this is to rewrite $g(x)$ defined as a function in parts. Recall that if $x > 6$, then $|x - 6| = x - 6$. If $x < 6$, then $|x - 6| = -(x - 6) = -x + 6$. Therefore,

$$g(x) = \begin{cases} x(-x + 6) = -x^2 + 6x & \text{for } x < 6 \\ x(x - 6) = x^2 - 6x & \text{for } x > 6. \end{cases}$$

We have two separate formulas for $g(x)$ depending on whether $x < 6$ or whether $x > 6$. Actually either formula is correct when $x = 6$ since $f(6) = 0$. The formula for $g(x)$ when $x < 6$ is $g(x) = -x^2 + 6x$. Thus $g'(x) = -2x + 6$ when $x < 6$. The formula for $g(x)$ when $x > 6$ is $g(x) = x^2 - 6x$. Thus $g'(x) = 2x - 6$ when $x > 6$. In order to find $g'(6)$ we need to use the definition of derivative. Using the definition we would see that $g'(6)$ does not exist.

$$f'(x) = \begin{cases} -2x + 6 & x < 6 \\ DNE & x = 6 \\ 2x - 6 & x > 6. \end{cases}$$

Note that $f'(x) = 0$ only for $x = 3$. If $x < 6$, then $g'(x) = -2x + 6$. It follows that $g'(3) = 0$. Let c denote a number such that $c < 3$, then $g'(c) = -2c + 6 = -2(c - 3) > 0$. Applying Theorem 1 we have that $g(x)$ is increasing for $x < 3$. Let c denote a number such that $3 < c < 6$, then $g'(c) = -2(c - 3) < 0$. Applying Theorem 2 we have that $g(x)$ is decreasing for $3 < x < 6$. If $x > 6$, then $g(x) = x^2 - 6x$ and $g'(x) = 2x - 6 = 2(x - 3) > 0$. Applying Theorem 1 we have that $g(x)$ is increasing for $x > 6$.

Example 10. Consider the function defined in parts by

$$f(x) = \begin{cases} x^2 + 4 & x < 0 \\ -3x + 4 & 0 \le x \le 3 \\ x^2 - 6x + 4 & x > 3. \end{cases}$$

For what values of x is this function increasing and for what values of x is it decreasing?

Solution. First, let us be clear about what this definition of a function says. If $x < 0$, say $x = -2$, then we substitute into $x^2 + 4$ to find the value of the function, that is, $f(-2)^2 + 4 = 8$. If $0 \le x \le 3$, say $x = 2$, then we substitute into $-3x + 4$ to find the value of the function, that is, $f(2) = -3(2) + 4 = -2$. If $x > 3$, say $x = 6$, then we substitute into $x^2 - 6x + 4$ to find the value of the function, that is, $f(6) = (6)^2 - 6(6) + 4 = 4$. We need to find the derivative of $f(x)$. If $x < 0$, then $f(x)$ is defined on both sides of x by $f(x) = x^2 + 4$, and so $f'(x) = 2x$. If x is in the open interval $0 < x < 3$, then $f(x)$ is defined on both sides of x by $f(x) = -3x + 4$ and so $f'(x) = -3$. If $x > 3$, then $f(x)$ is defined on both sides of x by $f(x) = x^2 - 6x + 4$ and so $f'(x) = 2x - 6$. Thus

$$f'(x) = \begin{cases} 2x & x < 0 \\ -3 & 0 < x < 3 \\ 2x - 6 & x > 3. \end{cases}$$

Note that $2(0) \ne -3$ and $-3 \ne 2(3) - 6$. This indicates that $f'(0) = DNE$ and $f'(3) = DNE$. If we absolutely need to we can prove that $f'(0) = DNE$ and $f(3) = DNE$ using the definition of derivative. Therefore,

$$f'(x) = \begin{cases} 2x & x < 0 \\ DNE & x = 0 \\ -3 & 0 < x < 3 \\ DNE & x = 3 \\ 2x - 6 & x > 3. \end{cases}$$

Since $f'(x) = 2x$ for $x < 0$ and $2x < 0$ for $x < 0$ it follows by Theorem 2 that $f(x)$ is decreasing for $x < 0$. Since $f'(x) = -3$ for $0 < x < 3$ and $-3 < 0$ it follows by Theorem 2 that $f(x)$ is decreasing for $0 < x < 3$. Since $f'(x) = 2x - 6$ for $x > 3$ and $3(x - 3) > 0$ for $x - 3 > 0$ it follows by Theorem 1 that $f(x)$ is increasing for $x > 3$.

Exercises

1. For what values of x are the following functions not defined:

a) $f(x) = \dfrac{x^3 + 8}{(2x - 5)(x + 4)}$
b) $f(x) = \ln |x - 3|$

c) $f(x) = e^{5x} + x^{1/2}$ (d) $f(x) = \sqrt{x^2 - 36}$

e) $f(x) = \sqrt{36 - x^2}$ (f) $f(x) = \tan(x/2)$.

2. Show by the method of Example 2 that the function $f(x) = x^2 + 5x$ is increasing at any number c such that $c > 0$.

3. Let $f(x) = x^3 - 6x^2 + 3x + 12$. Show using the method of Example 3 that the function $f(x)$ is decreasing at $c = 2$.

4. For what values of x is each of the following functions increasing? Use Theorem 1.

a) $f(x) = x^3 - 6x^2 - 36x$

b) $f(x) = xe^{-x/3}$

c) $f(x) = (4 + x^2)^{-1}$

d) $f(x) = (x - 4)^{2/3}$

5. For what values of x is each of the following functions increasing? For what values of x is each of the following functions decreasing?

a) $g(x) = (16 - x^2)^{1/3}$

b) $f(x) = x|x - 4|$

c) $f(x) = \begin{cases} x^2 - 2x & x \geq 3 \\ x & x < 3. \end{cases}$

d) $f(x) = \begin{cases} x^2 - 3x + 2 & x < 0 \\ x^2 + 2 & 0 \leq x \leq 3 \\ -x + 14 & x > 3. \end{cases}$

6. Sketch the graph of a function which satisfies all the following conditions:

a) $f'(x) > 0$ for $-3 < x < 5$

b) $f'(x) < 0$ for $x < -3$ and $x > 5$.

6355 First Derivative Test

First, let us review some facts about limits. In particular, let us be sure that we understand one sided limits. An ordinary two sided limit is written as

$$\lim_{x \to c} f(x).$$

In order to find the value of this limit we consider values of x as x gets closer and closer to c. In these two sided limits we must allow both values of x greater than c and values of x less than c. For example, in order to find

$$\lim_{x \to 2} \frac{\sqrt{x^2 - 1}}{x + 4}$$

we substitute values for x that are both less than 2 and values of x that are greater than 2. Since this function is continuous at $x = 2$, recall that we have a short cut method for finding the limit. We just substitute 2 for x in order to find the limit. We get

$$\lim_{x \to 2} \frac{\sqrt{x^2 - 1}}{x + 4} = \frac{\sqrt{3}}{6}.$$

What about this process changes when we have a one sided limit? There are two kinds of one sided limits. There are limits from the positive side (or from the right) and limits from the negative side (or from the left). The notation

$$\lim_{x \to c+} f(x)$$

indicates a one sided limit from the positive side. When finding this limit we let x get closer and closer to c, but we are restricted to values of x such that $x > c$.

Example 1. Find $\lim_{x \to 3+} [x^2 + \sqrt{x - 3}]$.

Solution. At first we may think that there is a problem because $x^2 + \sqrt{x - 3}$ is only defined for $x - 3 \geq 0$ or $x \geq 3$. This function is not defined for values of x on the negative side of 3. However, when finding $\lim_{x \to 3+}$ we only consider values of x such that $x > 3$. Since $x^2 + \sqrt{x - 3}$ is continuous function at $x = 3$ we have a short cut for finding the limit. We use the same short cut

141

that we used in order to find the value of a two sided limit for a continuous function. Thus, we are able to find the limit without having to actually substitute values of x with the values of x getting closer and closer to 3. We are able to find the value of the limit by just substituting 3 for x.

$$\lim_{x \to 3+} [x^2 + \sqrt{x - 3}] = 9.$$

Next look at limits from the left. The notation

$$\lim_{x \to c-} f(x)$$

indicates a one sided limit from the negative side. When finding this limit we let x get closer and closer to c, but are restricted to values of x such that $x < c$.

Example 2. Find $\lim_{x \to 4-} \sqrt{4 - x} + \sqrt{5 + x}$.

Solution. The function $\sqrt{4 - x}$ is only defined for $4 - x \geq 0$ or $x \leq 4$. The function $\sqrt{5 + x}$ is only defined for $x \geq -5$. The function $\sqrt{4 - x} + \sqrt{5 + x}$ is only defined for $-5 \leq x \leq 4$. But in order to find $\lim_{x \to 4-}$ we only substitute values of x less than 4 that get closer and closer to 4. We can do this because $\sqrt{4 - x} + \sqrt{5 + x}$ is defined for $-5 \leq x \leq 4$. Since $\sqrt{4 - x} + \sqrt{5 + x}$ is continuous we do not have to actually substitute values of x with the values of x getting closer and closer to 4. We can find the limit for limits from the positive side by just substituting 4 for x.

$$\lim_{x \to 4-} \sqrt{4 - x} + \sqrt{5 + x} = \sqrt{9} = 3.$$

Example 3. Find $\lim_{x \to 3+} [x^2(x - 3)]$.

Solution. Since the function $x^2(x - 3)$ is continuous, we can find the limit by substituting 3 for x.

$$\lim_{x \to 3+} x^2(x - 3) = 3^2(3 - 3) = 0.$$

The thing we want to notice about this example is the following. First, when finding the limit using the definition we substitute values for x which are greater than 3. This means that $x - 3 > 0$ or $x^2(x - 3) > 0$. Even though all the numbers $x^2(x - 3)$ are positive we still find that the limit is zero. However, when the numbers $x^2(x - 3)$ are all positive, the limit cannot be a negative number. In general suppose $g(x) > 0$ for $x > c$, it is still possible that $\lim_{x \to c+} g(x) = 0$. It is not possible for the value of the limit to be a negative number.

Definition of local maximum. Given a function $f(x)$ if there exists some number $d > 0$ such that $f(x) < f(c)$ for all x such that $0 < |x - c| < d$ and $x \neq c$, then we say that $f(x)$ has a local maximum value at $x = c$. Note that $d > 0$ is required but we do need to choose d sufficiently small.

This definition says that $f(x)$ has a local maximum at $x = c$ if given any value of x close to c the number $f(x)$ is less than the number $f(c)$.

Example 4. Let $f(x) = x^3 - 12x^2 + 45x$. Show that $f(x)$ has a local maximum value at $x = 3$.

Solution. We need to show that $f(x) < f(3)$ for all x such that $0 < |x - 3| < d$ for a sufficiently small number d. First,

$$f(3) = 3^3 - 12(3^2) + 45(3) = 54.$$

$$f(x) - f(3) = x^3 - 12x^2 + 45x - 54$$
$$= (x - 3)^2(x - 6).$$

For all x, $x \neq 3$, $(x - 3)^2 > 0$. For all x such that $|x - 3| < 2$, we have $-2 < x - 3 < 2$ or $-5 < x - 6 < -1$. Thus, if $|x - 3| < 2$, then $x - 6 < 0$ and so $(x - 3)^2(x - 6) < 0$. If $|x - 3| < 2$ and $x \neq 3$, then

$$f(x) - f(3) < 0$$
$$f(x) < f(3).$$

We have shown that $f(x)$ has a local maximum value at $x = 3$. Note that $d = 2$.

Definition of local minimum. Given a function $f(x)$ if there exists some number $d > 0$ such that $f(x) > f(c)$ for all x such that $0 < |x - c| < d$ and $x \neq c$ then we say that $f(x)$ has a local minimum value at $x = c$. Note that $d > 0$ is required but we do need to choose d sufficiently small.

This definition says that $f(x)$ has a local minimum value at $x = c$ if given any value of x close to c the number $f(x)$ is greater than the number $f(c)$.

Example 5. Let $f(x) = x^3 - 12x^2 + 45x$. Show that $f(x)$ has a local minimum value at $x = 5$.

Solution. We need to show that $f(x) > f(5)$ for all x such that $0 < |x-5| < d$ for a sufficiently small number d. First

$$f(5) = 5^3 - 12(5^2) + 45(5) = 50.$$

$$f(x) - f(5) = x^3 - 12x^2 + 45x - 50$$
$$= (x - 5)^2(x - 2).$$

For all x, $x \neq 5$, we have $(x - 5)^2 > 0$. For all x such that $|x - 5| < 1$ or $-1 < x - 5 < 1$ or $4 < x < 6$, we have $2 < x - 2 < 4$. Thus, if $|x - 5| < 1$ and $x \neq 5$, then $(x - 5)^2(x - 2) > 2(x - 5)^2 > 0$. If $|x - 5| < 1$ and $x \neq 5$, then

$$f(x) - f(5) > 0$$
$$f(x) > f(5).$$

We have shown that $f(x)$ has a local minimum value at $x = 5$. Note that we showed we could use the value $d = 1$.

Theorem 1. If $f'(c)$ exists and if $f(x)$ has a local maximum value at $x = c$, then $f'(c) = 0$.

Proof. When we say that $f'(c)$ exists we are saying that the two sided limit

$$\lim_{x \to c} \frac{f(x) - f(c)}{x - c}$$

exits. This says that both the one sided limits

$$\lim_{x \to c+} \frac{f(x) - f(c)}{x - c} \quad \text{and} \quad \lim_{x \to c-} \frac{f(x) - f(c)}{x - c}$$

exist and are both equal to $f'(c)$. Since $f(x)$ has a local maximum at $x = c$, the definition of local maximum tells us that if $x < c$ and x is near c, then

$f(x) < f(c)$. This says that $f(x) - f(c)$ is equal to a negative number for x sufficiently near c. If $x < c$, and $|x - c|$ is small, then

$$\frac{f(x) - f(c)}{x - c} = \frac{\text{negative}}{\text{negative}} = \text{positive}.$$

This means that

$$f'(c) = \lim_{x \to c-} \frac{f(x) - f(c)}{x - c}$$

is the limit of some positive numbers. This means that the limit is either a positive number or zero.

The definition of local maximum tells us that if we consider values of x such that $x > c$, then we still have $f(x) < f(c)$. Since $x - c > 0$, it must be true that

$$\frac{f(x) - f(c)}{x - c} = \frac{\text{negative}}{\text{positive}} = \text{negative},$$

for x near c. This means that

$$f'(c) = \lim_{x \to c+} \frac{f(x) - f(c)}{x - c}$$

is the limit of negative numbers and so is either a negative number or zero. When considering $x < c$, we conclude $f'(c)$ is positive or zero. When considering $x > c$, we conclude that $f'(c)$ is negative or zero. The only possible conclusion is that $f'(c) = 0$. Since the two sided limit exists it must equal the only possible common possible value of the two one sided limits.

Theorem 2. If $f'(c)$ exists and $f(x)$ has a local minimum value of $x = c$, then $f'(c) = 0$.

Proof. The proof is very similar to the proof of Theorem 1 and will be left to the reader.

In general when discussing local maxima and local minima for a function we will always assume that the function is continuous for all values of x for which the function is defined. If we allow functions to have "jump discontinuities" (be piecewise continuous) then the discussion of how to locate max and min points requires several more theorems. We choose not to introduce this complication at this time.

Theorem 3. The First Derivative Test. If $f'(x) > 0$ for $x < c$ and $f'(x) < 0$ for $x > c$ whenever $|x - c|$ is a sufficiently small number, then $f(x)$ has a local maximum value at $x = c$.

Proof. We are given that $f(x)$ is a function such that $f'(x) > 0$ for $x < c$. This implies that $f(x)$ is increasing for $x < c$ and x near c. We are given that $f(x)$ is such that $f'(x) < 0$ for $x > c$. This implies that $f(x)$ is decreasing for $x > c$ and x near c. The function $f(x)$ is increasing for $x < c$ and decreasing for $x > c$. This says that $f(x)$ has a local maximum at $x = c$. This can be shown more directly using a discussion similar to the discussion used to prove Theorem 1.

Theorem 4. The First Derivative Test. If $f'(x) < 0$ for $x < c$ and $f'(x) > 0$ for $x > c$ for x sufficiently close to c, then $f(x)$ has a local minimum value at $x = c$.

Proof. We are given that $f(x)$ is a function such that $f'(x) < 0$ for $x < c$. This implies that $f(x)$ is decreasing for $x < c$ and x near c. We are given that $f(x)$ is such that $f'(x) > 0$ for $x > c$. This implies that $f(x)$ is increasing for $x > c$ and x near c. The function $f(x)$ is decreasing for $x < c$ and increasing for $x > c$. This says that $f(x)$ has a local minimum at $x = c$.

Theorem 3 and Theorem 4 taken together are called the "First Derivative Test". We may use the term "First Derivative Test" when referring to either Theorem 3 or to Theorem 4.

Example 6. Let $f(x) = 5xe^{-x/5}$. Show that $f(x)$ has a local maximum value at $x = 5$.

Solution. We will use Theorem 3, the First Derivative Test. We need

$$f'(x) = 5[e^{-x/5} - (x/5)e^{-x/5}] = e^{-x/5}[5 - x].$$

If $x < 5$, then $x - 5 < 0$ or $5 - x > 0$. Note that $e^{-x/5} > 0$ for all x. If $x < 5$, then $f'(x) = e^{-x/5}[5 - x] > 0$. If $x > 5$, then $x - 5 > 0$ or $5 - x < 0$. If $x > 5$, then $f'(x) = e^{-x/5}[5 - x] < 0$. We have $f'(x) > 0$ for $x < 5$ and

146

$f'(x) < 0$ for $x > 5$. Applying the First Derivative Test we conclude that $f(x)$ has a local maximum value at $x = 5$.

Example 7. Let $f(x) = (1/3)x^3 - 4x^2 + 16x$. Does $f(x)$ have either a maximum value or a minimum value at $x = 4$?

Solution. First, $f'(x) = x^2 - 8x + 16 = (x - 4)^2$. It is clear that $f'(x) > 0$ for all x except $x = 4$. If $x < 4$, then $f'(x) > 0$. If $x > 4$, then $f'(x) > 0$. If $x < 4$, then $f(x)$ is an increasing function. If $x > 4$, then $f(x)$ is an increasing function. The function $f(x)$ is increasing at $x = 4$. The function $f(x) = (1/3)x^3 - 4x^2 + 16x$ has neither a maximum value nor a minimum value at $x = 4$.

Given a value of x, say $x = c$, Theorem 3 enables us to determine if $x = c$ is a point where $f(x)$ has a local maximum value. Theorem 4 enables us to determine if $x = c$ is a point where $f(x)$ has a local minimum value. These two theorems taken together are called "the First Derivative Test" for local maximum and local minimum. The First Derivative Test works in all cases when we are given the value of x to be tested. However, suppose we need to find the points that we need to test to determine if $f(x)$ has either a local maximum or a local minimum at these points. Suppose we have the problem: Find all values of x which give a local maximum value for $f(x) = x^3 - x^2 - 45x$. The solution of this problem involves two steps. The first step is to identify all values of x that can possibly give a maximum value or a minimum value of $f(x)$. The second step is to prove that certain of these points do indeed give a local maximum value of the function. The First Derivative Test enables us to do the second step. The question we must address is how do we do the first step. We look at Theorems 1 and 2 for help in answering this question. Theorem 1 and Theorem 2 are nice theorems but they are not really ideal. Taken together they say:

> If $f'(c)$ exists and either $f(x)$ has a local maximum value at $x = c$ or $f(x)$ has a local minimum value at $x = c$, then $f'(c) = 0$.

This says assuming $f'(x)$ exists if $x = c$ is either a local maximum or a local minimum then $f'(c) = 0$. We must be concerned with the possibility that other conditions besides local maximum and local minimum might also make $f'(c) = 0$. See Example 7. The function in Example 7 was such that

$f'(4) = 0$ but $f(x)$ did not have either a local maximum value or a local minimum value at $x = 4$.

Also note that all our discussion so far about how to determine if $f(x)$ has a maximum value at $x = c$ or a minimum value at $x = c$ starts with the assumption that $f'(c)$ exists. Suppose $f'(c)$ does not exist, is it possible that $f(x)$ has a local maximum value at $x = c$?

Example 8. Does $f(x) = (x - 4)^{2/3}$ have a local maximum or local minimum value at $x = 4$?

Solution. Note that $f'(x) = (2/3)(x - 4)^{-1/3}$ for $x \neq 4$. We use the definition of derivative to find $f'(4)$.

$$f'(4) = \lim_{x \to 4} \frac{(4 - x)^{2/3} - 0}{x - 4} = \lim_{x \to 4} \left[-(4 - x)^{-1/3}\right] = DNE.$$

This shows that $f'(4)$ does not exist. If $x < 4$, then $(x - 4) < 0$ and $(x - 4)^{-1/3} < 0$. If $x < 4$, then $f'(x) < 0$. If $x > 4$, then $(x - 4)^{-1/3} > 0$ and $f'(x) > 0$. Theorem 4 tells us that $f(x) = (x - 4)^{2/3}$ has a local minimum value at $x = 4$. For this example we have that $f(x)$ has a local minimum at $x = 4$, but $f'(4)$ does not exist. This tells us that if $f'(c)$ does not exist, then it is possible that $f(x)$ has a local maximum or a local minimum at $x = c$. Theorems 1 and 2 tell us that if $x = c$ is a local maximum or a local minimum and $f'(c)$ exists, then $f'(c) = 0$. Thus there are two kinds of values of x such that $f(x)$ may have a local maximum or a local minimum for these values of x. These are values of x, say $x = c$, such that $f'(c) = 0$ or $f'(c) = DNE$.

Definition. The number c is called a critical point (value) for the function $f(x)$ if either $f'(c)$ does not exist or $f'(c) = 0$.

Theorem 5. If $x = c$ is a local maximum or a local minimum value of $f(x)$, then c is a critical point of $f(x)$.

As stated above the maximum and minimum points are contained in the critical points. There are no maximum or minimum points that are not critical points. There may be some critical points that are not points that give either a maximum or minimum.

We encounter two situations where it is true that $f'(c) = DNE$. Suppose $f(x)$ is continuous at $x = c$ and $\lim\limits_{x \to c_-} f'(x)$ and $\lim\limits_{x \to c_+} f(x)$ exists but $\lim\limits_{x \to c_-} f'(x) \neq \lim\limits_{x \to c_+} f'(x)$, then $f'(x) = DNE$. If $f(x)$ is not continuous at $x = c$, then $f'(c) = DNE$. We do not discuss $f(x)$ such that $f(x)$ is not continuous at $x = c$, as part of the discussion of local maximum and local minimum values.

Example 9. Find all the values of x which give either a local maximum value or a local minimum value for $f(x) = x^3 - 3x^2 - 45x$.

Solution. Step 1. Find all the critical points for $f(x)$. Since $f'(x)$ is defined for all x, the list of critical points is just the points where $f'(x) = 0$.

$$f'(x) = 3x^2 - 6x - 45 = 3(x - 5)(x + 3).$$

It is clear that $f'(5) = 0$ and $f'(-3) = 0$. The critical points are $x = -3$ and $x = 5$. These are the only possible values of x where the function $f(x)$ can have a local maximum value or a local minimum value.

Step 2: which of these critical points give a maximum value for $f(x)$ and which give a minimum value for $f(x)$? We answer this question by applying the First Derivative Test. The First Derivative Test requires that we know the values of x for which $f'(x)$ is positive and the values of x for which $f'(x)$ is negative. Since $f'(-3) = 0$ start with a value of x less than -3 say $x = -4$, then $f'(-4) = 27$. Since $f'(5) = 0$ use a value of x between -3 and 5 say $x = 0$, then $f'(0) = -45$. Use a value of x greater than 5. Let $x = 6$, then $f'(6) = 27$. The only points where $f'(x) = 0$ are $x = -3$ and $x = 5$. Since $f'(x)$ is continuous this means that $f'(x)$ can only change sign at $x = -3$ and $x = 5$. Thus $f'(x) > 0$ for $x < 3$, $f'(x) < 0$ for $3 < x < 5$, and $f'(x) > 0$ for $x > 5$. We indicate the sign of $f'(x)$ on the following chart.

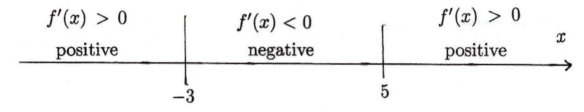

If $x < -3$, then $f'(x) > 0$ and if $x > -3$, then $f'(x) < 0$. Applying the First Derivative Test we conclude that $f(x)$ has a local maximum value at

$x = -3$. If $x < 5$, then $f'(x) < 0$ and if $x > 5$, then $f'(x) > 0$. Applying Theorem 4, the First Derivative Test, we conclude that $f(x)$ has a local minimum value at $x = 5$.

Given a function $f(x)$ suppose we have a graph of $y = f(x)$. It is possible to determine which values of x give a local maximum value of $f(x)$ and which values of x give a local minimum value of $f(x)$ by looking at the graph of $f(x)$. Suppose that $x = c$ is a value of x that gives a local maximum value of $f(x)$, then $f(x) < f(c)$ for all x near c. This means that the point with coordinates $(c, f(c))$ is locally the highest point on the graph. Given a graph we would first find the highest point on the graph and determine the coordinates of this point. Say the coordinates are (x_1, y_1). This means that $x_1 = c$ and $y_1 = f(c)$. The x value, $x = c = x_1$ is the x value which gives a local maximum value for $f(x)$. Suppose that $x = b$ is a value of x that gives a local minimum value of $f(x)$, then $f(x) > f(b)$ for all x near b. This means that the point with coordinates $(b, f(b))$ is locally the lowest point on the graph. Given a graph, we would first find the lowest point on the graph and determine the coordinates of this point. Say that the coordinates are (x_2, y_2). This means that $x_2 = b$ and $y_2 = f(b)$. The x value, $x = b = x_2$, is the x value which gives a local minimum value for $f(x)$. This minimum value is $f(b)$.

Example 10. The following is the graph of $y = f(x)$. For what values of x does the function $f(x)$ have a local maximum value and for what values of x does $f(x)$ have a local minimum value.

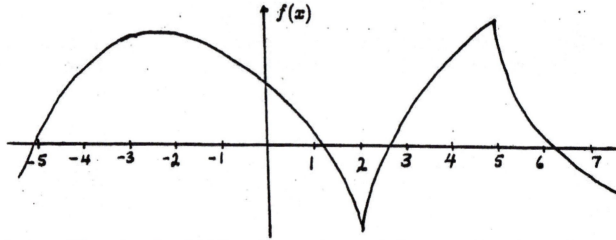

Solution. There is a local high point on the graph for $x = -3$ and $x = 5$. The function $f(x)$ has a local maximum value when $x = -3$ and when

$x = 5$. Since the scale on the y axis is not shown we can not determine the maximum values. There is a local low point on the graph when $x = 2$. The function $f(x)$ has a local minimum value when $x = 2$.

Example 11. Find all the values of x for which the function $f(x) = x|x-4|$ has either a local maximum value or a local minimum value.

Solution. Step 1. Note that $f(x)$ is continuous for all values of x. Find all the critical numbers for $f(x)$. We need to find $f'(x)$. The easiest way to do this is to break $f(x)$ up into two parts.

$$f(x) = x(x - 4) \quad \text{when } x > 4.$$
$$f'(x) = 2x - 4 \quad \text{when } x > 4.$$

Also

$$f(x) = -x(x - 4) \text{ when } x < 4$$
$$f'(x) = -2x + 4 \text{ when } x < 4.$$

Let us sketch a graph of $f'(x)$.

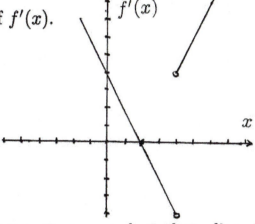

Suppose $f(x)$ is continuous at $x = c$, but that $\lim\limits_{x \to c-} f'(x) \neq \lim\limits_{x \to c+} f'(x)$, then $f'(c) = DNE$. For this example $\lim\limits_{x \to 4-} f'(x) = -4$ and $\lim\limits_{x \to 4+} f'(x) = 4$. It follows that $f'(4) = DNE$. Also $f'(2) = 0$. We can check this in $f'(x) = -2x + 4$ for $x < 4$. Substituting $x = 2$, we get $f'(2) = -2(2) + 4 = 0$. The critical points are $x = 2$ and $x = 4$. We get $x = 4$ is a critical point because $f'(4)$ does not exist. We can easily decide for what values of x $f'(x)$ is positive and for what values of x $f'(x)$ is negative by looking at the graph of $f'(x)$. For $x < 2$, we see $f'(x) > 0$ and for $2 < x < 4$ we see $f'(x) < 0$. Applying Theorem 3, the First Derivative Test, the function $f(x)$ has a local maximum at $x = 2$. Next, consider the other critical value

$x = 4$. For $2 < x < 4$, $f'(x) < 0$ and for $x > 4$, $f'(x) > 0$. Applying the First Derivative Test, Theorem 4, the function $f(x)$ has a local minimum value at $x = 4$.

Example 12. Let $f(x) = (x - 2)^{2/3} - (1/3)x$. For what values of x does this function have a local maximum value and for what values of x does it have a local minimum value.

Solution. The first step is to find the critical points for $f(x) = (x - 2)^{2/3} - (1/3)x$. Note that this function is defined for all values of x and is continuous.

$$f'(x) = (2/3)(x - 2)^{-1/3} - (1/3).$$

Note that $f'(2)$ is not defined. Thus, $x = 2$ is a critical point. Setting $f'(x) = 0$, we get

$$(2/3)(x - 2)^{-1/3} - (1/3) = 0$$
$$(x - 2)^{-1/3} = 1/2$$
$$(x - 2)^{1/3} = 2$$
$$(x - 2) = 8$$
$$x = 10.$$

Since $f'(10) = 0$, then $x = 10$ is a critical point. In order to apply the first derivative test we need to know the sign of $f'(x)$. If $x < 2$, then $x - 2 < 0$,

$$(x - 2)^{1/3} < 0,$$
$$(x - 2)^{-1/3} < 0,$$
$$(2/3)(x - 2)^{-1/3} - 1/3 < -1/3 < 0.$$

Thus, $f'(x) < 0$ when $x < 2$. On the other hand, if $2 < x < 10$, we have

$$0 < x - 2 < 8$$
$$0 < (x - 2)^{1/3} < 2$$
$$(x - 2)^{-1/3} > 1/2$$
$$2/3(x - 2)^{-1/3} > 1/3$$
$$2/3(x - 2)^{-1/3} - 1/3 > 0$$

Thus, $f'(x) > 0$ when $2 < x < 10$. Also note that $f'(3) = 1/3$.

If $x > 10$, then

$$x - 2 > 8, \ (x - 2)^{1/3} > 2, \ \text{and} \ (x - 2)^{-1/3} < 1/2.$$

It follows that $2/3(x - 2)^{-1/3} < 1/3$ and $2/3(x - 2)^{-1/3} - 1/3 < 0$. Thus, $f'(x) < 0$ when $x > 10$. We summarize this on the number line

$f'(x)$ negative	$f'(x)$ positive	$f'(x)$ negative

2   10

Applying the First Derivative Test (Theorem 4) at the critical point $x = 2$ we conclude that $f(x)$ has a local minimum value at $x = 2$. This minimum value is $f(2) = -(2/3)$. Applying the First Derivative Test (Theorem 3) at the critical point $x = 10$ we conclude that $f(x)$ has a local maximum value at $x = 10$. This maximum value is $f(10) = 2/3$.

Example 13. Suppose $f(x)$ is a function with the following properties:

(a) $f'(-1) = 0$ and $f'(3) = DNE$

(b) $f'(x) > 0$ whenever $-1 < x < 3$.

(c) $f'(x) < 0$ whenever either $x < -1$ or $x > 3$.

Find the critical points for $f(x)$ and determine which of these critical points give a local maximum value of $f(x)$ and which give a local minimum value of $f(x)$.

Solution. Note that problems of this type are shorter than problems where $f(x)$ is explicitly given. In these problems we do not have to do the first step which is to look at the function and determine for what values of x the derivative is positive and for which values the derivative is negative. These values are given. The only critical numbers are the values of x such that $f'(x) = 0$ and $f'(x) = DNE$. The critical numbers are -1 and 3.

$$f'(x) < 0 \qquad f'(x) > 0 \qquad f'(x) < 0$$

x

$-1 \qquad\qquad 3$

When $x < -1$, $f'(x) < 0$ and when $-1 < x < 3$, $f'(x) > 0$. Applying Theorem 4, the First Derivative Test, we conclude that $f(x)$ has a local minimum value at $x = -1$. When $-1 < x < 3$, $f'(x) > 0$ and when $x > 3$, $f'(x) < 0$. Applying Theorem 3, we conclude that $f(x)$ has a local maximum value at $x = 3$.

Exercises

1. Show that $f(x) = x^3 - 3x^2 - 24x + 15$ has a local minimum value at $x = 4$ using the method of Example 5.

2. Show that $f(x) = x^3 - 3x^2 - 24x + 15$ has a local maximum value at $x = -2$ using the method of Example 4.

3. Find the critical numbers for each of the following functions. Second, classify each critical number as either a point where the function $f(x)$ has a local maximum or a local minimum or as a point where the function has neither a local maximum nor a local minimum using the First Derivative Test.

    (a) $f(x) = x^3 - 6x^2 - 15x + 8$

    (b) $f(x) = \dfrac{x}{4 + x^2}$

    (c) $f(x) = x^2 e^{-x/3}$. The function $f(x)$ has local max at $x = 2$.

    (d) $f(x) = x^3 - 9x^2 + 27x + 45$

4. Find the critical numbers for each of the following functions. Second, classify each of the critical numbers as either a point where the function $f(x)$ has a local maximum or as a point where the function has a local minimum, or as a point where the function has neither a local maximum nor a local minimum using the First Derivative Test.

    (a) $f(x) = x^{5/3} - 5x^{2/3}$

    (b) $f(x) = x|x - 6|$

154

(c) $f(x) = (x - 4)^{2/3} - (2/9)x$

(d) $f(x) = 12(x - 1)^{1/3} - x$

(e) $f(x) = \begin{cases} x^2 - 4 & x \le 3 \\ -x + 8 & x > 3. \end{cases}$

(f) $f(x) = \dfrac{2x^2 + 22x - 24}{x^2 + 2x}$

5. The function $f(x)$ satisfies the conditions:

   (a) $f'(-2) = 0$ and $f'(3) = DNE$

   (b) $f'(x) > 0$ when $x < -2$ and when $x > 3$

   (c) $f'(x) < 0$ when $-2 < x < 3$.

What are the critical points for $f(x)$? Use the First Derivative Test to determine if these critical numbers are values of x where the function $f(x)$ has a local maximum or where the functions $f(x)$ has a local minimum, or neither.

6. The function $f(x)$ satisfies the conditions:

   (a) $f'(-3) = DNE$, $f'(1) = 0$, and $f'(5) = 0$.

   (b) $f'(x) > 0$ when $x < -3$ and when $x > 5$.

   (c) $f'(x) < 0$ when $-3 < x < 1$ and when $1 < x < 5$.

What are the critical points for $f(x)$? Use the First Derivative Test to determine if these critical numbers are values of x where the function $f(x)$ has a local maximum or where the function $f(x)$ has a local minimum, or neither.

6357 Concavity

We have discussed the idea of increasing and decreasing for a function. When discussing increasing and decreasing of a function it is especially helpful to look at the graph of the function. When sketching the graph of a function we realize there is another idea which can be very helpful. This idea is to use concavity.

Definition of concave up. If for all values of x near $x = c$ the graph of $y = f(x)$ is above the graph of the tangent line to the curve at $x = c$, then we say that the graph of the function $f(x)$ is **concave up** at $x = c$.

We often shorten "the graph of the function $f(x)$ is concave up at $x = c$" to "the function $f(x)$ is concave up at $x = c$". Also note that if $f'(c)$ does not exist (no tangent line), then we do not define the term concave up at $x = c$. Let $y = T_c(x)$ denote the equation of the tangent line to the curve $y = f(x)$ at $x = c$. If $f(x) > T_c(x)$ for all values of x such that x is near c but $x \neq c$, then $f(x)$ is concave up at $x = c$. Note that both the curve $y = f(x)$ and the line $y = T_c(x)$ pass through the point $(c, f(c))$. This says that $f(c) = T_c(c)$.

Example 1. Show that the graph of the function $f(x) = x^3 - 8x^2 + 12x$ is concave up at $x = 5$.

Solution. We need to find the equation of the tangent line to the curve $y = x^3 - 8x^2 + 12x$ at $x = 5$. We have $f'(x) = 3x^2 - 16x + 12$ and $f'(5) = 3(25) - 16(5) + 12 = 7$. Also $f(5) = 5^3 - 8(5^2) + 12(5) = -15$. The equation of the tangent line is

$$y - (-15) = 7(x - 5)$$
$$y = T_5(x) = 7x - 50.$$

The graph of $f(x) = x^3 - 8x^2 + 12x$ is concave up at $x = 5$ if $f(x) > T_5(x)$ for all x near $x = 5$ and $x \neq 5$. The graph of $f(x)$ is concave up if

$$x^3 - 8x^2 + 12x > 7x - 50,$$

for all values of x near $x = 5$ but $x \neq 5$. This is the same as

$$x^3 - 8x^2 + 5x + 50 > 0.$$

Factoring we see that this is the same as

$$(x + 2)(x - 5)^2 > 0.$$

If $x > -2$, then $(x + 2) > 0$. If $x \neq 5$, then $(x - 5)^2 > 0$. If $x > -2$ and $x \neq 5$, then $(x + 2)(x - 5)^2 > 0$. Thus, $f(x) > T_5(x)$ for $x > -2$. The graph of the function $y = f(x)$ is above the graph of the tangent line $y = T_5(x)$ for values of x such that $x > -2$. This certainly includes all values of x near $x = 5$. The function $f(x)$ is concave up at $x = 5$.

Theorem 1. If $f''(c) > 0$, then the function $f(x)$ is concave up at $x = c$.

Proof. Note that we assume $f''(c)$ exists and so $f'(c)$ exists. Let $y = T_c(x)$ denote the equation of the tangent line to the graph of $y = f(x)$ at the point $x = c$. According to the definition of concave up if $f(x) > T_c(x)$ for all x near c and $x \neq c$, then $f(x)$ is concave up at $x = c$. Define a new function $g(x)$ by

$$g(x) = f(x) - T_c(x).$$

If $g(x) > 0$ for all x near c but $x \neq c$, then $f(x)$ is concave up at $x = c$. Recall that the equation of the tangent line to $y = f(x)$ at $x = c$ is

$$y = T_c(x) = f(c) + f'(c)(x - c).$$

Therefore,

$$g(x) = f(x) - f(c) - f'(c)(x - c)$$
$$g'(x) = f'(x) - f'(c)$$
$$g''(x) = f''(x)$$

Substituting $x = c$, we get

$$g(c) = f(c) - f(c) - f'(c)(c - c) = 0$$
$$g'(c) = f'(c) - f'(c) = 0$$
$$g''(c) = f''(c).$$

Since it is given that $f''(c) > 0$, then $g''(c) > 0$. Since $g''(c) > 0$, the function $g'(x)$ is an increasing function at $x = c$. (If the derivative of $g'(x)$ is positive, then the function $g'(x)$ is increasing). Since $g'(c) = 0$ and $g'(x)$

is increasing at $x = c$, it follows that $g'(x) < 0$ for $x < c$ and $g'(x) > 0$ for $x > c$. The First Derivative Test says: if $g'(x) < 0$ for $x < c$ and $g'(x) > 0$ for $x > c$, then $g(x)$ has a local minimum value at $x = c$. Thus by the first derivative test the function $g(x)$ has a local minimum value at $x = c$. Since $g(c) = 0$ and $g(x)$ has a local minimum value at $x = c$, it follows that for all values of x near $x = c$ but $x \neq c$ it must be true that $g(x) > 0$. This says

$$g(x) = f(x) - T_c(x) > 0 \text{ or } f(x) > T_c(x)$$

for all x near $x = c$ and $x \neq c$. The graph of the function $f(x)$ is above the graph of its tangent line $T_c(x)$ for x near c. The function $f(x)$ is concave up at $x = c$.

Definition of concave down. If for values of x near $x = c$ the graph of $y = f(x)$ is below the graph of the tangent line to the curve at $x = c$ but $x \neq c$, then we say that the function $f(x)$ is **concave down** at $x = c$.

Let $y = T_c(x)$ denote the equation of the tangent line to the curve $y = f(x)$ at $x = c$. If $f(x) < T_c(x)$ for all values of x near $x = c$, then $f(x)$ is concave down at $x = c$. Note that both the curve $y = f(x)$ and the tangent line $y = T_c(x)$ pass through the point $(c, f(c))$.

Example 2. Show that $f(x) = x^3 - 8x^2 + 12x$ is concave down at $x = 1$.

Solution. We need to find the equation of the tangent line to the graph of $y = x^3 - 8x^2 + 12x$ at $x = 1$. First, $f(1) = 5$. We have $f'(x) = 3x^2 - 16x + 12$ and so $f'(1) = -1$. The equation of the tangent line is

$$y - (5) = -1(x - 1)$$
$$y = T_1(x) = -x + 6.$$

The graph of $f(x) = x^3 - 8x^2 + 12x$ is concave down at $x = 1$ if $f(x) < T_1(x)$ for all x near $x = 1$ but $x \neq 1$. The graph of $f(x)$ is concave down at $x = 1$ if

$$x^3 - 8x^2 + 12x < -x + 6,$$

for all values of x near $x = 1$ but $x \neq 1$. This is the same as

$$x^3 - 8x^2 + 13x - 6 < 0.$$

Factoring we see that this is the same as

$$(x-1)^2(x-6) < 0.$$

If $x < 6$, then $x - 6 < 0$. If $x \neq 1$, then $(x-1)^2 > 0$. If $x < 6$ and $x \neq 1$, then $(x-1)^2(x-6) < 0$. If $x < 6$ and $x \neq 1$, then $f(x) - T_1(x) < 0$ or $f(x) < T_1(x)$. If $x < 6$ and $x \neq 1$, the graph of the function $y = f(x)$ is below the graph of the tangent line. Note that $x < 6$ includes all x near $x = 1$. The graph of the function $f(x)$ is concave down at $x = 1$.

When we say that the graph of $y = f(x)$ at a point is concave up, then as part of the proof of Theorem 1 we showed that $f'(x)$ is increasing at that point. This means that as you move along the curve by increasing x the slope of the tangent line increases and so the tangent line to the curve turns counterclockwise. When we say that the graph of $y = f(x)$ at a point is concave down this means that $f'(x)$ is decreasing at that point. This means that as you move along the curve by increasing x the slope of the tangent line decreases and so the tangent line to the curve turns clockwise.

Theorem 2. If $f''(c) < 0$, then the function $f(x)$ is concave down at $x = c$.

The proof of Theorem 2 is very similar to the proof of Theorem 1 and is left to the reader.

The function $f(x) = (x-8)^4$ is concave up at $x = 8$ and $f''(8) = 0$. On the other hand, $f(x) = (x-8)^3$ is not concave up and is not concave down at $x = 8$ and $f''(8) = 0$. If $f''(c) = 0$, then $f(x)$ can be concave up at $x = c$ or $f(x)$ can be concave down at $x = c$, or $f(x)$ can be neither concave up nor concave down at $x = c$. If we just know that $f''(c) = 0$, we can not draw any conclusion about the concavity of $f(x)$ at $x = c$.

Let $g(x) = (x-8)^{4/3}$. We have $g'(x) = (4/3)(x-8)^{1/3}$ and so $g'(8) = 0$. The equation of the tangent line is $T(x) \equiv 0$. The expression $g(x) - T(x) = (x-8)^{4/3} \geq 0$. This shows that $g(x) > T(x)$ for $x \neq 8$. The function $g(x) = (x-8)^{4/3}$ is concave up at $x = 8$. On the other hand, $g''(x) = (4/9)(x-8)^{-2/3}$ and so $g''(8)$ does not exist. A function can be concave up at $x = c$ or concave down at $x = c$ and $g'(c)$ exist, but $g''(c)$ does not exist.

Example 3. Let $f(x) = x(x-3)^2$. Show that the graph of $f(x)$ is concave up for any value of x such that $2 < x < 10$.

Solution. We want to use Theorem 1 which says: if $f''(c) > 0$, then $f(x)$ is concave up at $x = c$. We need the second derivative.

$$f'(x) = 3x^2 - 12x + 9$$
$$f''(x) = 6x - 12$$

Clearly, $f''(x) > 0$ for all x such that $6x - 12 > 0$ or $x > 2$. We conclude that $f(x)$ is concave up for values of x such that $x > 2$ which includes $2 < x < 10$.

Example 4. Show that the graph of $f(x) = \dfrac{x}{x^2 + 12}$ is concave down for any value of x such that $0 < x < 6$.

Solution. We want to use Theorem 2. We need the second derivative. Use the quotient rule to differentiate and simplify, we get

$$f'(x) = \frac{-x^2 + 12}{(x^2 + 12)^2}$$

Use the quotient rule again and simplify, we get

$$f''(x) = \frac{2x(x^2 - 36)}{(x^2 + 12)^3}.$$

Note that the denominator of this fraction is larger than 1728 for all values of x. The denominator is always positive. The sign of the fraction is determined by the sign of the numerator. For $0 < x < 6$, we have $x^2 - 36 < 0$ and so $x(x^2 - 36) < 0$. It follows that if $0 < x < 6$, then $f''(x) < 0$. Theorem 2 then tells us if $f''(x) < 0$, then $f(x)$ is concave down. It follows that if $0 < x < 6$, then the graph of $y = f(x)$ is concave down.

Example 5. Let $f(x) = \frac{\sin x}{2 + \cos x}$. Show that the graph of $f(x)$ is concave down at any value of x such that $0 < x < \pi$.

Solution. Note that $f(x)$ is defined for all values of x. We want to use Theorem 1. We need the second derivative. We use the quotient rule.

$$f'(x) = \frac{(2 + \cos x)(\cos x) - \sin x(-\sin x)}{(2 + \cos x)^2}$$

Simplifying we get

$$f'(x) = \frac{2\cos x + 1}{(2 + \cos x)^2}$$

Again we use the quotient rule for finding the derivative of a quotient.

$$f''(x) = \frac{(2 + \cos x)^2(-2\sin x) - (2\cos x + 1)(2)(2 + \cos x)(-\sin x)}{(2 + \cos x)^4}$$

Simplifying we get

$$f''(x) = \frac{2\sin x(\cos x - 1)}{(2 + \cos x)^3}$$

Note that $f''(x)$ is defined for all x. Also note that $f''(0) = 0$, and $f''(\pi) = 0$. Note that $(2 + \cos x) > 0$. If $0 < x < \pi$, then $\sin x > 0$. If $0 < x < \pi$, then $\cos x < 1$ and $\cos x - 1 < 0$. It follows that if $0 < x < \pi$ then $(\sin x)(\cos x - 1) < 0$. If $0 < x < \pi$, then $f''(x) < 0$. Applying Theorem 2 we conclude: if $0 < x < \pi$, then the graph of $f(x)$ is concave down.

We should be able to show concavity when sketching a graph. The first thing to remember is that a straight line segment is neither concave up nor is it concave down. The direction of concavity shows the direction a curve is bending. The direction in which a curve is bending can be found by looking at the slope of the tangent lines as x increases. If the slopes are increasing, then the curve is concave up. If the slopes are decreasing, then the curve is concave down. Let us look at a couple of examples that make the shape of the curve clear. First, consider the function $f(x) = x^2$. Since $f''(x) = 2$ for all x, this function is concave up for all values of x. A sketch of the graph of $y = x^2$ is

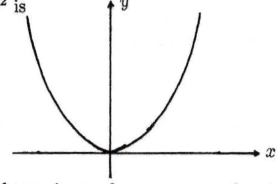

If a function is decreasing and concave up, then its graph takes the shape of the left half of this curve. If a function is increasing and concave up, then its graph takes the shape of the right half of this curve.

161

Second, consider $g(x) = -x^2$. Since $g''(x) = -2$, the function $g(x) = -x^2$ is concave down for all values of x. A sketch of the graph of $y = -x^2$ is

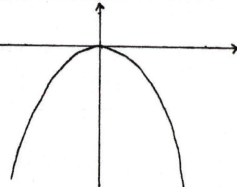

If a function is increasing and concave down, then its graph takes the shape of the left half of this curve. If a function is decreasing and concave down, then its graph takes the shape of the right half of this curve.

Example 6. Suppose $f(x)$ is a continuous function that has all the following properties, then sketch the graph of $y = f(x)$ for $-5 \le x \le 6$.

(a) $f'(-3) = DNE$, $f'(0) = 0$, $f'(2) > 0$, and $f'(4) = 0$, $f''(-3) = DNE$, $f''(0) > 0$ $f''(2) = 0$, $f''(4) = 0$.

(b) $f'(x) > 0$ for $-5 < x < -3$, for $0 < x < 2$, $2 < x < 4$, and for $4 < x < 6$.

(c) $f'(x) < 0$ for $-3 < x < 0$.

(d) $f''(x) > 0$ for $-3 < x < 2$ and for $4 < x < 6$.

(e) $f''(x) < 0$ for $-5 < x < -3$ and for $2 < x < 4$.

Solution. We need to locate all points where $f'(x) = 0$ or $f'(x) = DNE$. Also all points where $f''(x)$ changes sign. The second derivative, $f''(x)$, changes sign at $x = -3$, $x = 2$, and $x = 4$. Plot all these values on a x number line.

$$f''(x) < 0 \qquad f''(x) > 0 \qquad f''(x) < 0 \quad f''(x) > 0$$
$$f'(x) > 0 \quad f'(x) < 0 \qquad f'(x) > 0 \qquad f'(x) > 0 \qquad x$$

$$-5 \qquad -3 \qquad 0 \qquad 2 \qquad 4 \qquad 6$$

We need to label each of the intervals determined by these division points. Label each interval as an interval where $f'(x)$ is positive or as an interval where $f'(x)$ is negative. Label each interval as an interval where $f''(x)$ is positive or as an interval where $f''(x)$ is negative.

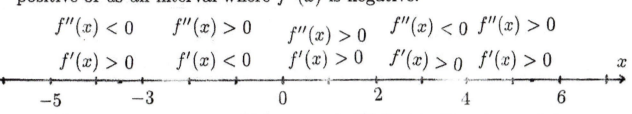

$$f''(x) < 0 \qquad f''(x) > 0 \qquad f''(x) > 0 \qquad f''(x) < 0 \quad f''(x) > 0$$
$$f'(x) > 0 \qquad f'(x) < 0 \qquad f'(x) > 0 \qquad f'(x) > 0 \quad f'(x) > 0 \qquad\qquad x$$

If $-5 < x < -3$, then $f'(x) > 0$ and $f''(x) < 0$. The shape of the curve for these values of x is

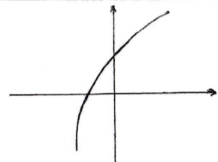

If $-3 < x < 0$, then $f'(x) < 0$ and $f''(x) > 0$. The shape of the curve for these values of x is

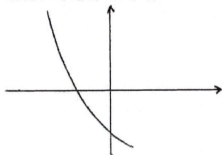

If $0 < x < 2$, then $f'(x) > 0$ and $f'(x) > 0$. The shape of the curve on this interval is

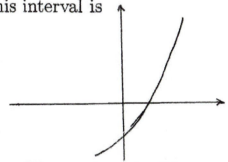

We continue in this manner. Using this information we sketch a continuous curve. Since we are given no values of $f(x)$, we have a lot of choice about how far to go up and down on the y axis.

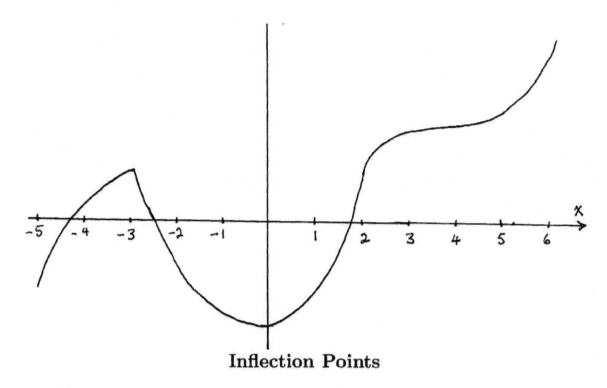

Inflection Points

Definition. Suppose $f(x)$ is a function which is continuous at $x = c$. The point with x coordinate $x = c$, and y coordinate $y = f(c)$ is called an **inflection point** of the graph of $y = f(x)$ if the graph changes concavity at $x = c$.

Note that $x = c$ can be an inflection point of the graph of $y = f(x)$ even through $f'(c)$ does not exist. However, in order to define concavity at a point the derivative must exist at that point.

Theorem 3. The Inflection Point Theorem. The point $x = c$ is an inflection point for $f(x)$ if either of the following is true.

a) $f''(x) > 0$ for $x < c$ and $f''(x) < 0$ for $x > c$.

b) $f''(x) < 0$ for $x < c$ and $f''(x) > 0$ for $x > c$.

Theorem 4. If $f''(c)$ exists and the graph of $f(x)$ has an inflection point at $x = c$, then $f''(c) = 0$.

Theorem 5. If $x = c$ is an inflection point of the graph of $y = f(x)$, then either $f''(c) = 0$ or $f''(c) = DNE$.

Given a value of x, say $x = c$, Theorem 3, the Inflection Point Theorem, enables us to determine if $x = c$ is an inflection point of $f(x)$. However,

164

suppose we want to find all the inflection points of $f(x)$. We do not have a perfect theorem. Theorem 5 tells us that if $x = c$ is an inflection point, then either $f''(c) = 0$ or $f''(c) = DNE$. This indicates that we must test all values of c such that $f''(c) = 0$ or $f''(c) = DNE$ using Theorem 3, the Inflection Point Theorem, to determine which points are actually inflection points.

Example 6. Find the inflection points if any for $f(x) = (x - 2)^{8/3}$.

Solution. First, note that $f(x) = [(x - 2)^{1/3}]^8$ is defined for all values of x and $f(x) \geq 0$ for all values of x. The first two derivatives are

$$f'(x) = (8/3)(x - 2)^{5/3}$$
$$f''(x) = (40/9)(x - 2)^{2/3}.$$

By Theorem 5 in order for $x = c$ to be an inflection point we must have either $f''(c) = 0$ or $f''(c) = DNE$. Note that $f''(2) = 0$ and $f''(x)$ is defined for all values of x. There is only one value to check. We should check to see if the graph of $y = f(x)$ has an inflection point at $x = 2$. Note that $(x - 2)^{2/3} \geq 0$ for all x. The second derivative $f''(x) = (40/9)(x-2)^{2/3}$ does not change sign at $x = 2$. There is no inflection point at $x = 2$. The graph of $y = (x - 2)^{8/3}$ has no inflection points. This is interesting. We have $f'(2) = 0$ and $f''(2) = 0$, but $x = 2$ is not an inflection point. Just for fun let us look at the first derivative. If $x < 2$, then $(x-2)^{5/3} < 0$. If $x > 2$, then $(x - 2)^{5/3} > 0$. If $x < 2$, then $f'(x) < 0$ and if $x > 2$, then $f'(x) > 0$. The First Derivative Test tells us that $f(x)$ has a local minimum at $x = 2$.

Example 7. Find the inflection points if any for $f(x) = (x - 2)^{1/3}$.

Solution. First, note that $f(x) = (x - 2)^{1/3}$ is defined for all values of x. The first two derivatives are

$$f'(x) = (1/3)(x - 2)^{-2/3}$$
$$f''(x) = (-2/9)(x - 2)^{-5/3}.$$

We clearly see that $f''(x)$ is not zero for any value of x and that $f''(2)$ does not exist. Following Theorem 5, we should check to see if the graph of $y = f(x)$ has an inflection point at $(2, 0)$. If $x < 2$, then $x - 2 < 0$

and $(x-2)^{-5/3} < 0$. If $x > 2$, then $x - 2 > 0$ and $(x-2)^{-5/3} > 0$. If $x < 2$, then $f''(x) < 0$ and if $x > 2$, then $f''(x) > 0$. The second derivative changes sign at $x = 2$. Applying the Inflection Point Theorem the function $f(x) = (x-2)^{1/3}$ has an inflection point at $x = 2$. There are no other inflection points. Note that since $f'(2) = DNE$ the graph of $y = f(x)$ has no tangent line at $x = 2$ and so the concavity of the graph of $f(x)$ is not defined at $x = 2$. Look at the first derivative. Clearly $f'(x) > 0$ for all values of x except $x = 2$. The first derivative does not change sign at $x = 2$. The function $f(x)$ has neither a maximum value nor a minimum value at $x = 2$.

If a short section of the graph of $f(x)$ near $x = c$ is a line segment, then the graph of $f(x)$ is not concave at $x = c$.

Exercises

1. Show that $f(x) = x^3 - 6x^2 + 5x$ is concave up at $x = 4$ using the same method that was used in Example 1. Show that $f(x) = x^3 - 6x^2 + 5x$ is concave down at $x = -1$ using the same method that was used in Example 2.

2. Show that $f(x) = x^3 - 9x^2 + 10x + 8$ is concave up for all x such that $x > 3$. Use Theorem 1.

3. Show that the graph of $f(x) = \frac{x}{x^2+9}$ is concave down for all x such that $0 < x < \sqrt{27}$. Use Theorem 2.

4. For what values of x is the graph of the function $f(x)$ concave up, for what values of x is the graph of $f(x)$ concave down, and for what values is it not concave?

$$f(x) = \begin{cases} -x & x < 0 \\ x^2 - 5x & 0 \le x \le 5 \\ -x^2 + 15x - 50 & x > 5. \end{cases}$$

Note that $f(x)$ is continuous.

5. Suppose $f(x)$ is a continuous function for all values of x that has all the following properties. For what values of x is the function decreasing? For what value of $f(x)$ is concave up? You may want to sketch a graph.

a) $f'(-3) = DNE$, $f'(-1) = DNE$, $f'(2) = 0$, $f'(4) = DNE$, $f'(6) = 0$, $f''(-3) = DNE$, $f''(-1) = DNE$, $f''(2) > 0$, $f''(4) = DNE$, and $f''(6) < 0$.

b) $f'(x) > 0$ for $-3 < x < -1$, for $2 < x < 4$, and for $4 < x < 6$.

c) $f'(x) < 0$ for $x < -3$, for $-1 < x < 2$ and for $x > 6$.

d) $f''(x) > 0$ for $-3 < x < -1$, for $-1 < x < 2$, and for $2 < x < 4$.

e) $f''(x) < 0$ for $x < -3$, for $4 < x < 6$, and for $x > 6$.

6. Suppose $f(x)$ is a continuous function for all values of x that has all the following properties. Note that $f'(x)$ exists for all values of x except $x = -2$ and $x = 1$. For what values of x is concavity not defined? For what values of x is $f(x)$ increasing? For what values of x is $f(x)$ concave up? Sketch the graph for $y = f(x)$ for $-4 < x < 8$.

a) $f'(-2) = DNE$, $f'(1) = DNE$, $f'(5) = 0$, $f''(-2) = DNE$, and $f''(1) = DNE$, $f''(5) < 0$.

b) $f'(x) > 0$ for $x < -2$ and for $1 < x < 5$.

c) $f'(x) < 0$ for $-2 < x < 1$ and for $x > 5$.

d) $f''(x) > 0$ for $x < -2$ and for $-2 < x < 1$

e) $f''(x) < 0$ for $1 < x < 5$ and for $x > 5$.

7. For what values of x is the graph of the function $f(x) = x^{5/3} - 5x^{2/3}$ increasing and for what values is the graph decreasing? For what values of x is the graph concave up and for what values of x is the graph concave down? Sketch the graph. Note that $f(0) = 0$. Concavity is not defined at $x = 0$ since $f'(0) = DNE$.

8. Find all the inflection points for $f(x) = x^4 - 12x^3 - 15x$.

9. Find all the inflection points if any for $f(x) = (x - 3)^4$.

10. Find all the inflection points if any for $f(x) = (x - 3)^{5/3}$.

11. Let $f(x) = 9(x - 2)^{8/3} - 80x^2$ then $f''(x) = 40(x - 2)^{2/3} - 160$ and $f''(10) = 0$. Show that $x = 10$ is an inflection point on the graph of $f(x)$.

12. Suppose $f(x)$ is a continuous function for all real numbers x with all the following properties. Also $f'(x)$ is continuous for all x except for $x = -3$ and $x = 5$.

a) $f'(-3) = DNE$, $f'(2) = 0$, $f'(5) = DNE$, $f''(-3) = DNE$, $f''(2) = 0$, $f''(5) = DNE$.

b) $f'(x) > 0$ for $x < -3$, for $-3 < x < 2$, and for $2 < x < 5$.

c) $f'(x) < 0$ for $x > 5$.

d) $f''(x) > 0$ for $x < -3$ and for $2 < x < 5$.

e) $f''(x) < 0$ for $-3 < x < 2$ and for $x > 5$.

Find the x coordinate of all the inflection points of the graph of $f(x)$.

13. Let $f(x) = x^{7/3} - 28x^{1/3}$. For what values of x is the function $f(x)$ increasing? For what values of x is the graph of $f(x)$ concave up?

6359 Second Derivative Test

We have already discussed a good test for determining if a function $f(x)$ has a local maximum value or a local minimum value at $x = c$. This test is the First Derivative Test. If the first derivative changes sign from plus to minus, then we have a local maximum value. If the first derivative changes sign from minus to plus, then we have a local minimum value. We also have another test which we can use to determine if $f(x)$ has a local maximum value or a local minimum value at $x = c$. This test is called the second derivative test. It is easier to apply than the First Derivative Test, but it has the disadvantage. The second derivative test fails to resolve the question in some cases.

Theorem 1. The Second Derivative Test. If $f'(c) = 0$ and $f''(c) > 0$, then $f(x)$ has a local minimum value at $x = c$.

Proof. Suppose that $f(x)$ is a function such that $f'(c) = 0$ and $f''(c) > 0$. Since $f'(c) = 0$, the tangent line to the curve $y = f(x)$ at $x = c$ has a slope of zero, and so is parallel to the x axis. The equation of this tangent line $y = f(c)$. Since $f''(c) > 0$, the graph of $y = f(x)$ is concave up at $x = c$. Since the graph is concave up the graph of $y = f(x)$ is above the graph of the tangent line, $y = f(c)$, for x near c. Therefore, $f(x) > f(c)$ for x near c and $x \neq c$ which says that $f(x)$ has a local minimum value at $x = c$.

Theorem 2. The Second Derivative Test. If $f'(c) = 0$ and $f''(c) < 0$, then $f(x)$ has a local maximum at $x = c$.

The proof of Theorem 2 is similar to the proof of Theorem 1. Theorem 1 and Theorem 2 taken together are called the "Second Derivative Test". We may use the term "Second Derivative Test" to refer to either of these theorems. Note that if $f''(c) = DNE$ or if $f''(c) = 0$, then the Second Derivative Test does not lead to a conclusion at $x = c$. The Second Derivative Test fails.

Example 1. Find the values of x where the function $f(x) = e^x \cos x$ has a local maximum and where $f(x)$ has a local minimum.

Solution. First, find the critical points for $f(x)$. These are the points where $f(x)$ might have a local maximum or a local minimum value.

$$f'(x) = e^x \cos x - e^x \sin x = e^x(\cos x - \sin x).$$

Since $e^x \neq 0$ setting $f'(x) = 0$, gives $\cos x - \sin x = 0$, or $\tan x = 1$. The critical points are

$$x = \pi/4 + 2n\pi \text{ and } x = 5\pi/4 + 2n\pi.$$

Let us just test $x = \pi/4$ and $x = 5\pi/4$ using the Second Derivative Test. The other points are very similiar.

$$f''(x) = -2e^x \sin x.$$
$$f''(\pi/4) = -2e^{\pi/4} \sin(\pi/4) = -\sqrt{2}e^{\pi/4} < 0.$$

By Theorem 1, the Second Derivative Test, the function $f(x) = e^x \cos x$ has a local maximum value at $x = \pi/4$. The maximum value is $f(\pi/4) = e^{\pi/4} \cos \pi/4 = (\sqrt{2}/2)e^{\pi/4} \simeq 1.5509$. The function also has a local maximum value at $x = \pi/4 + 2n\pi$.

Let us test the critical point $x = 5\pi/4$.

$$f''(5\pi/4) = -2e^{5\pi/4} \sin(5\pi/4) = \sqrt{2}e^{5\pi/4} > 0.$$

By Theorem 2, the Second Derivative Test, the function $f(x) = e^x \cos x$ has a local minimum value at $x = 5\pi/4$. This minimum value is $f(5\pi/4) = e^{5\pi/4} \cos(5\pi/4) = -(\sqrt{2}/2)e^{5\pi/4} \simeq -35.8885$. The function $f(x) = e^x \cos x$ also has a local minimum value at $x = 5\pi/4 + 2n\pi$.

Example 2. Consider the function

$$f(x) = x^4 - 12x^3 + 48x^2 - 64x + 10.$$

Show that $x = 1$ and $x = 4$ are critical points. Test these critical points for local maximum and local minimum using the second derivative test.

Solution. First, $f'(x) = 4x^3 - 36x^2 + 96x - 64$

$$f'(1) = 4 - 36 + 96 - 64 = 0$$
$$f'(4) = 4(4^3) - 36(4^2) + 96(4) - 64 = 0.$$

Since $f'(1) = 0$ and $f'(4) = 0$ this shows that both $x = 1$ and $x = 4$ are critical points.

$$f''(x) = 12x^2 - 72x + 96$$
$$f''(1) = 12 - 72 + 96 = 36 > 0.$$

170

Applying the Second Derivative Test, Theorem 1, since $f''(1) > 0$, we conclude that $f(x)$ has a local minimum at $x = 1$. The value of this minimum is $f(1) = -17$. Next, let us consider the critical value $x = 4$.

$$f''(4) = 12(4^2) - 72(4) + 96 = 0.$$

Since $f''(4) = 0$ neither Theorem 1 nor Theorem 2 can be applied. Since $f''(4) = 0$ the Second Derivative Test does not apply or fails. The Second Derivative Test fails to give any information. The First Derivative Test can be applied at this critical point. The First Derivative Test would tell us that $f(x)$ has neither a local maximum nor a local minimum at $x = 4$.

Exercises

1. Find the values of x where the function $f(x) = e^{\sqrt{3}x} \cos x$ has a local maximum and where $f(x)$ has a local minimum. The critical points are $x = (\pi/3) + n\pi$, $n = 0, 1, 2, 3, \cdots$, $n = -1, -2, -3, \cdots$.

2. Consider the function $f(x) = x^4 - 4x^3 + 16x + 10$. Show that $x = -1$ and $x = 2$ are critical points. Test these critical points for local maximum and local minimum values using the Second Derivative Test.

3. Consider the function $f(x) = (x - 2)^{4/3}$. Find the critical points. Test the critical points to see if they give a local maximum or a local minimum value using the Second Derivative Test. Second, test the critical points to see if they give a local maximum or a local minimum value using the First Derivative Test.

4. Consider the function $f(x) = x^2 e^{-3x}$. Find the critical points. Test these critical points to see if they give a local maximum or a local minimum value using the Second Derivative Test.

5. Consider the function $f(x) = 2x^3 - 9x^2 - 108x + 100$. Find the critical points. Test these critical points to see if they give a local maximum or a local minimum value using the Second Derivative Test.

6. Find the critical points of the following function. Test these critical points using the Second Derivative Test to determine if they give a local maximum value or a local minimum value

$$f(x) = \frac{x}{x^2 + 9}.$$

7. Suppose that $f(x)$ is a continuous function with all the following properties.

(a) $f'(-3) = DNE$, $f'(1) = 0$, $f'(4) = DNE$, $f''(-3) = DNE$, $f''(1) = 0$, and $f''(4) = DNE$.

(b) $f'(x) > 0$ for $-3 < x < 1$ and for $x > 4$. Also $f'(x) < 0$ for $x < -3$ and for $1 < x < 4$.

(c) $f''(x) > 0$ for $x > 4$. Also $f''(x) < 0$ for $x < -3$, for $-3 < x < 1$, and for $1 < x < 4$.

Find the critical points. Test the critical points to see if they give a local maximum value of $f(x)$ or a local minimum value of $f(x)$ using the Second Derivative Test. Second, test the critical points to see if they give a local maximum value of $f(x)$ or a local minimum value of $f(x)$ using the First Derivative Test.

8. Suppose that $f(x)$ is a continuous function with all the following properties

(a) $f'(-3) = 0$, $f'(1) = DNE$, $f'(4) = 0$, $f''(-3) = 0$, $f''(1) = DNE$, and $f''(4) = 0$.

(b) $f'(x) > 0$ for $x < -3$, for $1 < x < 4$, and for $x > 4$. Also $f'(x) < 0$ for $-3 < x < 1$.

(c) $f''(x) > 0$ for $x > 4$. Also $f''(x) < 0$ for $x < -3$, for $-3 < x < 1$, and for $1 < x < 4$.

Find the critical points for $f(x)$. Test the critical points using the Second Derivative Test to see if they give a local maximum value of $f(x)$ or a local minimum value of $f(x)$. Second, test the critical points using the First Derivative Test to see if they give a local maximum value of $f(x)$ or a local minimum value of $f(x)$ or neither.

6360 Maximum and Minimum at Boundary Points

In our previous discussion about finding the local maximum and minimum values of a function we have only considered the following situation. We said that the function $f(x)$ has a local maximum value at $x = c$ if $f(x) < f(c)$ all values of x such that $|x - c| < d$ for a sufficiently small d. It was required that we be able to check both values of x less than c and values of x greater than c. We did not consider the question unless $f(x)$ was defined for all x such that $|x - c|$ was sufficiently small. We required that the function $f(x)$ be defined for values of x on both sides of c. We now wish to define the terms "local maximum" and "local minimum" when the function $f(x)$ is defined only on one side of c. Consider the function $f(x) = (64 - x^2)^{3/2}$. This function is naturally defined for $-8 \le x \le 8$. This function can not be defined using this formula for either $x < -8$ or $x > 8$. If $c = -8$ or $c = 8$, then the function is defined on only one side of c. All the values of x taken collectively for which a function is defined is called the domain of the function. The domain of this function is $-8 \le x \le 8$. The numbers -8 and 8 are called end points of the interval (domain) of definition of this function. The numbers -8 and 8 are also called boundary points of the domain of definition of this function.

In our previous discussion of local maximum and local minimum we had a theorem which was easy to apply. The theorem said "if $f'(x)$ changes sign from positive to negative at $x = c$, then $f(x)$ has a local maximum value at $x = c$". It would be nice to have a theorem like this that involves $f'(x)$ to cover the case of end points. The problem with this idea is that we have never defined the derivative of a function at an end point. Recall our definition of the derivative of $f(x)$ at $x = c$.

$$f'(c) = \lim_{x \to c} \frac{f(x) - f(c)}{x - c}.$$

This is a two sided limit and requires that $f(x)$ be defined both for $x < c$ and for $x > c$. Recall that we have also defined and worked with one sided limits. This means that we have a relatively simple way to define the derivative at an end point.

Definition 1. If $f(x)$ is only defined for $x \ge a$, then

$$f'(a) = \lim_{x \to a+} \frac{f(x) - f(a)}{x - a}.$$

173

This definition says that if $f(x) = (25 - x^2)^{3/2}$ for $-5 \le x \le 5$, then

$$f'(-5) = \lim_{x \to (-5)+} \frac{(25 - x^2)^{3/2} - 0}{x - (-5)} = \lim_{x \to (-5)+} \left[-(x - 5)\sqrt{25 - x^2} \right] = 0.$$

Another function. This definition says that if $g(x) = (16 - x^2)^{1/2}$ for $-4 \le x \le 4$, then

$$g'(-4) = \lim_{x \to (-4)+} \frac{(16 - x^2)^{1/2} - 0}{x - (-4)} = \lim_{x \to (-4)+} \left[\frac{4 - x}{4 + x} \right]^{1/2} = +\infty.$$

The derivative $g'(-4)$ does not exist.

Suppose $f(x)$ is only defined for $x \ge a$. Whenever $\lim_{x \to a+} \frac{f(x) - f(a)}{x - a} = +\infty$ we use the shorthand notation $f'(a) = +\infty$. Whenever $\lim_{x \to a+} \frac{f(x) - f(a)}{x - a} = -\infty$ we use the shorthand notation $f'(a) = -\infty$. In both cases we are actually saying that $f'(a)$ does not exist and so may write $f'(a) = DNE$.

Definition 2. If $f(x)$ is defined only for $x \le b$, then

$$f'(b) = \lim_{x \to b-} \frac{f(x) - f(b)}{x - b}.$$

This definition says that if $f(x) = (25 - x^2)^{3/2}$ for $-5 \le x \le 5$, then

$$f'(5) = \lim_{x \to 5-} \frac{(25 - x^2)^{3/2} - 0}{x - 5} = \lim_{x \to 5-} \left[-(x + 5)\sqrt{25 - x^2} \right] = 0.$$

Another function. This definition says that if $g(x) = (16 - x^2)^{1/2}$ for $-4 \le x \le 4$, then

$$g'(4) = \lim_{x \to 4-} \frac{(16 - x^2)^{1/2} - 0}{x - 4} = \lim_{x \to 4-} \left[-\left(\frac{x + 4}{4 - x} \right)^{1/2} \right] = -\infty.$$

The derivative $g'(4)$ does not exist. Suppose $f(x)$ is only defined for $x \le b$. Whenever $\lim_{x \to b-} \frac{f(x) - f(b)}{x - b} = +\infty$ we use the shorthand notation $f'(b) =$

174

$+\infty$ and whenever $\lim\limits_{x\to b-} \dfrac{f(x)-f(b)}{x-b} = -\infty$ we use the shorthand notation $f'(b) = -\infty$. In both cases we are actually saying that $f'(b)$ does not exist and so may write $f'(b) = DNE$.

Suppose that $f(x)$ is continuous for $x \geq a$ and is only defined for $x \geq a$. For a function only defined for $x \geq a$ we are only able to find the derivative $f'(x)$ using the usual rules for differentiation for $x > a$. However, in many cases the formula or algebraic expression that we get for $f'(x)$ also has a value or is defined for $x = a$. Denote this value by L, that is, L is the number we get when we substitute $x = a$ into the algebraic expression (formula) for $f'(x)$. Define the function $g(x)$ by

$$g(x) = \begin{cases} f'(x) & x > a \\ L & x = a. \end{cases}$$

In almost all cases the function $g(x)$ is continuous for $x \geq a$. Note that $g(x)$ is given by the same algebraic expression as is $f'(x)$. Let us consider the situation where $g(x)$ is continuous at $x = a$. Later we will use the Mean Value Theorem for Derivatives to prove: If $f(x)$ is continuous for $x \geq a$ with x near a and $f'(x)$ is continuous for $x > a$ with x near a, then

$$\frac{f(x)-f(a)}{x-a} = f'(\theta)$$

for some number θ such that $a < \theta < x$. It follows that

$$f'(a) = \lim_{x\to a+} \frac{f(x)-f(a)}{x-a} = \lim_{x\to a+} f'(\theta) = \lim_{x\to a+} g(\theta) = L.$$

Note that if $\lim\limits_{x\to a+} f'(\theta) = DNE$, then $f'(a) = DNE$, we state these results as the following theorem

Theorem 1. Suppose $f(x)$ is defined for $x \geq a$ and the derivative $f'(x)$ is given by an explicit formula involving elementary functions and that the formula (algebraic expression) for $f'(x)$ defines a continuous function for $x \geq a$, then the one sided derivative of $f(x)$ at $x = a$ is given by $f'(a)$. This says that we can use the same formula (algebraic expression) to find $f'(a)$ that we used to find $f'(x)$ for $x > a$. Also if substituting $x = a$ into the formula (algebraic expression) for $f'(x)$ is not defined, then $f'(a) = DNE$.

Example 1. Suppose $f(x) = (x-2)^{5/2} + x^2$ for $x \geq 2$, then $f'(x) = (5/2)(x-2)^{3/2} + 2x$ for $x > 2$. This formula (algebraic expression) for $f'(x)$ defines a function which is continuous for $x \geq 2$ and so by Theorem 1 gives the value of $f'(2)$, that is, $f'(2) = 4$.

Example 2. Suppose $g(x) = (x-3)^{1/2} - x^3$ for $x \geq 3$, then using the usual rules for differentiation we find $g'(x) = (1/2)(x-3)^{-1/2} - 3x^2$ for $x > 3$. Substituting $x = 3$ into this algebraic expression for $g'(x)$ we need to divide by $(3-3)^{1/2}$ which is zero. It follows from Theorem 1 that $g'(3) = DNE$. we could, of course, show that $g'(3) = DNE$ by using the definition of derivative directly that is, $\lim_{x \to 3+} [(x-3)^{-1/2} - (x^2 + 3x + 9)]$.

Theorem 2. Suppose $f(x)$ is defined for $x \leq b$ and the derivative of $f'(x)$ is given by an explicit function (algebraic expression) involving elementary functions and that this formula (algebraic expression) defines a continuous function for $x \leq b$, then the one sided derivative of $f(x)$ at $x = b$ is given by $f'(b)$. This says that we can use the same formula to find $f'(b)$ that we used to find $f'(x)$ for $x < b$. Also if substituting $x = b$ into the formula (algebraic expression) for $f'(x)$ is not defined, then $f'(b) = DNE$.

Example 3. Suppose $f(x) = (16 - x^2)^{3/2} + 5x$ for $-4 \leq x \leq 4$, then using the usual rules $f'(x) = (-3x)(16 - x^2)^{1/2} + 5$ for $-4 < x < 4$. This expression for $f'(x)$ defines a function which is continuous for $-4 \leq x \leq 4$. Applying Theorem 2 this formula also gives the value of $f'(4)$, that is, $f'(4) = 5$.

Example 4. Suppose $g(x) = (25 - x^2)^{1/2} - x^3$ for $-5 \leq x \leq 5$. Note that $g(x)$ is continuous for $-5 \leq x \leq 5$. Using the usual rules of differentiation we find $g'(x) = (-x)(25 - x^2)^{-1/2} - 3x^2$ for $-5 < x < 5$. Substituting $x = 5$ into the expression for $g(x)$ we get undefined since $(5-5)^{-1/2}$ is undefined. It follows from Theorem 2 that $g'(5) = DNE$.

Definition 3. Suppose $f(x)$ is only defined for $x \geq a$, then a is called an "end point" or "boundary point" for the domain of definition of $f(x)$.

Definition 4. Suppose $f(x)$ is only defined for $x \geq a$. We say that $f(x)$ has a local maximum at $x = a$ if $f(x) < f(a)$ for all x such that $a < x < a + d$ for a sufficiently small d.

Definition 5. Suppose $f(x)$ is only defined for $x \geq a$. We say that $f(x)$ has a local minimum at $x = a$ if $f(x) > f(a)$ for all x such that $a < x < a + d$ for a sufficiently small d.

Theorem 3 (Left End Point Theorem). Suppose $f(x)$ is only defined for $x \geq a$. The function $f(x)$ has a local maximum value at $x = a$ if either $f'(a) < 0$ or $f'(a) = -\infty$. The function $f(x)$ has a local minimum value at $x = a$ if either $f'(a) > 0$ or $f'(a) = +\infty$.

Proof. Consider the case when $f'(a) < 0$. Since

$$f'(a) = \lim_{x \to a+} \frac{f(x) - f(a)}{x - a}$$

and $f(x)$ is continuous. This says that

$$\lim_{x \to a+} \frac{f(x) - f(a)}{x - a} < 0.$$

It follows that for a small enough value of d we have

$$\frac{f(x) - f(a)}{x - a} < 0$$

whenever $a < x < a + d$. Since $x - a > 0$, it follows that $f(x) - f(a) < 0$ or $f(x) < f(a)$ whenever $a < x < a + d$. This says that $f(x)$ has a local maximum value at $x = a$.

The other cases of the theorem are proved in a very similar way.

Example 5. Let $f(x) = (16 - x^2)^{3/2} + x^2$ for $-4 \geq x \leq 4$. Show that $f(x)$ has a local maximum at the end point $x = -4$.

Solution. Using the usual rules of differentiation $f'(x) = (-3x)(16 - x^2)^{1/2} + 2x$ for $-4 < x < 4$. Since $f'(x)$ is given by an algebraic expression which is continuous at $x = -4$, we have $f'(-4) = -8$. Applying the Left End Point Theorem we conclude that $f(x)$ has a local maximum value at $x = -4$. We know that $f'(x)$ is continuous for $x \geq -4$ because it is a combination of elementary functions and to find $f'(4)$ we do not have to divide by zero.

Example 6. Let $g(x) = 3x - (16 - x^2)^{1/2}$ for $-4 \leq x \leq 4$. Show that $f(x)$ has a local maximum value at the end point $x = -4$.

Solution. We have $g'(x) = 3 + x(16 - x^2)^{-1/2}$. Since $g'(x)$ is given by a formula and so continuous for $x > -4$ and $\lim\limits_{x \to (-4)+} [3 + x(16 - x^2)^{-1/2}] = -\infty$, we have $g'(-4) = -\infty$. Applying the Left End Point Theorem we conclude that $g(x)$ has a local maximum value at $x = -4$.

Example 7. Let $f(x) = 2(x - 4)^{3/2} + x^2$ for $x \geq 4$. Show that $f(x)$ has a local minimum value at the end point $x = 4$.

Solution. Using the usual rules for differentiation $f'(x) = 3(x-4)^{1/2} + 2x$ for $x > 4$. Since $f'(x)$ is given by an expression that is defined and continuous for $x \geq 4$, we have $f'(4) = 8$. Applying the Left End Point Theorem we conclude that $f(x)$ has a local minimum value at $x = 4$, since $f'(4) > 0$.

Recall we only discuss maximum and minimum problems for functions that are continuous. This means we assume $f(x)$ is continuous for $x \geq a$. When we write $f'(a)$ we assume that $f'(a)$ exists as explained by Theorem 1. That is, the formula (expression) $f'(x)$ is continuous for $x \geq a$. When we say $f'(a) = +\infty$ we mean that $f'(a) = DNE$ and $\lim_{x \to a+} f(x) = +\infty$.

Remark. In the Left End Point Theorem we can replace the statement "the function $f(x)$ has a local maximum value at $x = a$ if $f'(a) = -\infty$" with the more inclusive also true statement "the function $f(x)$ has a local maximum value at $x = a$ if $f(a)$ is defined and $f'(x) < 0$ for $a < x < a + d$ for sufficiently small d". However, this statement which covers more cases also makes the discussion of examples and problems much longer.

Definition 6. Suppose $f(x)$ is only defined for $x \leq b$. We say that $f(x)$ has a local maximum value at $x = b$ if $f(x) < f(b)$ for all x such that $b - d < x < b$ for a sufficiently small value of d with $d > 0$.

Definition 7. Suppose $f(x)$ is only defined for $x \leq b$. We say that $f(x)$ has a local minimum value at $x = b$ if $f(x) > f(b)$ for all x such that $b - d < x < b$ for a sufficiently small value of d.

Theorem 4 (Right End Point Theorem). Suppose $f(x)$ is only defined for $x \leq b$. The function $f(x)$ has a local maximum value at $x = b$ if either $f'(b) > 0$ or $f'(b) = +\infty$. The function $f(x)$ has a local minimum value at $x = b$ if either $f'(b) < 0$ or $f'(b) = -\infty$.

Proof. Consider the case when $f'(b) = +\infty$. Since

$$f'(b) = \lim_{x \to b-} \frac{f(x) - f(b)}{x - b}$$

this says that

$$\lim_{x \to b-} \frac{f(x) - f(b)}{x - b} = +\infty.$$

It follows that for a small enough value of d, we have

$$\frac{f(x) - f(b)}{x - b} > 0,$$

whenever $b - d < x < b$. Since $x - b < 0$, it follows that $f(x) - f(b) < 0$ or $f(x) < f(b)$ whenever $b - d < x < b$. This says that $f(x)$ has a local maximum value at $x = b$.

The other cases of the theorem are proved in a very similar way.

Example 8. Let $f(x) = 2(16 - x^2)^{3/2} + x^2$ for $-4 \le x \le 4$. Show that $f(x)$ has a local maximum value at $x = 4$.

Solution. Using the usual rules of differentiation we get $f'(x) = (-3x)(16 - x^2)^{1/2} + 2x$ for $-4 < x < 4$. Since $f'(x)$ is given by a formula that is defined and continuous for $x \le 4$, we have $f'(4) = 8$. The Right End Point Theorem says that if $f(x)$ is only defined for $x \le 4$ and $f'(4) > 0$, then $f(x)$ has a local maximum value at $x = 4$. We conclude that $f(x)$ has a local maximum value at $x = 4$.

Example 9. Let $g(x) = x^2 + \sqrt{16 - x^2}$ for $-4 \le x \le 4$. Show that $g(x)$ has a local minimum value at $x = 4$.

Solution. We have $g'(x) = 2x - x(16 - x^2)^{-1/2}$. Since $g'(x)$ is given by an expression which is continuous for $-4 < x < 4$ and $\lim_{x \to 4-} [2x - x(16 - x^2)^{-1/2}] = -\infty$, we have $g'(4) = -\infty$. Applying the Right End Point Theorem we conclude that $g(x)$ has a local minimum value at $x = 4$.

Example 10. Let $f(x) = 2(5 - x)^{3/2} - 3x^2 + 8$ for $x \le 5$. Show that $f(x)$ has a local minimum value at $x = 5$.

Solution. Using the usual rules for differentiation we get $f'(x) = -3(5 - x)^{1/2} - 6x$ for $x < 5$. Since the formula for $f(x)$ defines a continuous function for $x \le 5$, we have $f'(5) = -30$. Applying The Right End Point Theorem since $f'(5) < 0$ we conclude that $f(x)$ has a local minimum value at $x = 5$.

Example 11. Let $f(x) = 4(x - 6)^{1/2}$ for $x \ge 6$. Show that $f(x)$ has a local maximum value at $x = 6$.

Solution. Using the usual rules $f'(x) = 2(x - 6)^{-1/2}$ for $x > 6$. Note that $f'(6) = +\infty$. Applying the Left End Point Theorem we conclude that $f(x)$ has a local maximum value at $x = 6$.

The statement of the Left End Point Theorem and the Right End Point Theorem are similar, but are different in some important details. For this reason we need something to help us remember exactly what the theorems says. A good memory aid to help us remember is to draw a simple graphic such as the following. The vertical lines indicate the end points and the slanting lines the direction of the graph.

Left End Point Theorem
$f' < 0$ local maximum

Right End Point Theorem
$f' > 0$ local maximum

Left End Point Theorem
$f' > 0$ local minimum

Right End Point Theorem
$f' < 0$ local minimum

Suppose we have the function $f(x) = \frac{x}{(x-2)^2}$ defined for $x \ne 2$, and want to find the values of x which give a local maximum value or a local minimum value of the function We note that $f(x)$ is not defined when $x = 2$ and that $\lim_{x \to 2} \dfrac{x}{(x-2)^2} = +\infty$, that is, $\lim_{x \to 2} \dfrac{x}{(x-2)^2}$ does not exist. We may be tempted to say that $f(x)$ has a local maximum value at $x = 2$, but that is not correct. The function is not defined when $x = 2$. Since the function is not defined when $x = 2$, it does not have any kind of value when $x = 2$. Note that $f'(2)$ also does not exist. Infinity is not a number.

Example 12. Find all the points where the function $f(x) = (16 - x^2)^{-1/2}$, $-4 < x < 4$, has a local maximum value or a local minimum value.

Solution. Note that $f(x)$ is not defined for $x \leq -4$ or for $x \geq 4$. If $-4 < c < 4$, then $f(x)$ is defined for $x < c$ and x close to c. Also $f(x)$ is defined for $x > c$ and x close to c. This function $f(x)$ is defined on both sides of c. This means that we can use the first and second derivative tests to test for local maximum and local minimum values.

$$f'(x) = -\frac{1}{2}(16 - x^2)^{-3/2}(-2x) = \frac{x}{(16 - x^2)^{3/2}} \text{ for } -4 < x < 4.$$

We find the critical points by solving $f'(x) = 0$. The solution is easy. The only critical point is $x = 0$. What about end points? There are no end points. We might at first think that the right hand derivative at $x = 4$ is given by $f'(4) = -\infty$, but this is not the case. The function $f(x) = (16 - x^2)^{-1/2}$ is not defined for $x = 4$. The derivative $f'(4)$ is not defined. We can not write $f'(4) = -\infty$ because this means that

$$\lim_{x \to 4-} \frac{f(x) - f(4)}{x - 4} = -\infty.$$

This cannot be true because $f(4)$ is not defined. We do not have a maximum value or a minimum value of $f(x)$ at $x = 4$ because $f(4)$ is not defined. The value $x = 4$ is not a value of x being considered. For values of x near $x = 0$, we have $(16 - x^2)^{3/2} > 0$. If $x < 0$, then $f'(x) < 0$. If $x > 0$, then $f'(x) > 0$. The first derivative test tells us that $f(x)$ has a local minimum value at $x = 0$.

When discussing the ideas of local maximum value and local minimum value we have assumed that the function under discussion was continuous. The reason for this is that in this course we have spent almost no time discussing functions that are not continuous. To be clear, given $x = c$, if $f(x)$ is not continuous at $x = c$, then we will not discuss the problem of deciding if the function has a local maximum value or a local minimum value at $x = c$. This question is usually decided by considering the graph of $f(x)$ for values of x near $x = c$. Otherwise we must develop a whole list of new theorems to cover the discontinuous case.

In a lot of situations we have found that $\lim_{x \to c} f(x) = DNE$. In the past there have been two cases for which $\lim_{x \to c} f(x) = DNE$. We can express these two cases as $\lim\limits_{x \to c} f(x) = +\infty$ and $\lim\limits_{x \to c} f(x) = -\infty$. Since these two cases are the only ways we have had in the past such that the limit does not exist we might begin to think these two cases are the only possibilities. In our future work we will indeed mostly discuss $\lim\limits_{x \to c} f(x)$ does not exist in the case where $|f(x)| = +\infty$. However, there is another important case. This is the case where $\lim\limits_{x \to c-} f(x) \neq \lim\limits_{x \to c+} f(x)$ but both limits exist. In future we will consider a few examples of this case. In the long run this is really the most interesting way for $\lim\limits_{x \to c} f(x)$ to not exist. As a final note let us consider the limit

$$\lim_{x \to 0} \sin(\frac{1}{x}).$$

Note that $\lim\limits_{x \to 0} \sin(\frac{1}{x}) = DNE$. Also note that it is not true that

$\lim\limits_{x \to 0} \sin(\frac{1}{x}) = +\infty$ and it is not true that $\lim\limits_{x \to 0} \sin(\frac{1}{x}) = -\infty$. As x approaches zero the function $\sin(\frac{1}{x})$ oscillates. It will continue to be true in our future work that we will only find the two cases $\lim\limits_{x \to c} f(x) = +\infty$ and $\lim\limits_{x \to c} f(x) = -\infty$.

Exercises

1. Let $f(x) = (x + 2)^{3/2} + 5x^2$ for $x \geq -2$. Show that $f(x)$ has a local maximum value at the boundary point $x = -2$.

2. Let $f(x) = (25 - x^2)^{3/2} - x^2$ for $-5 < x < 5$. Show that $f(x)$ has a local minimum value at $x = -5$.

3. Let $f(x) = (25 - x^2)^{1/2} - 2x^2$ for $-5 \leq x \leq 5$. Does the function $f(x)$ have a local maximum value or a local minimum value at the end point $x = 5$?

4. Let $f(x) = (4 - x)^{5/2} + x^3$ for $x \leq 4$. Show that $f(x)$ has a local maximum value at the boundary point $x = 4$.

5. Let $f(x) = (x^2 - 16)^{3/2} + x^{1/2}$ for $x \geq 4$. Show that $f(x)$ has a local minimum value at the boundary point $x = 4$.

6. Let $f(x) = (5 + 4x - x^2)^{3/2} - 4x$ for $-1 \leq x \leq 5$. Show that $f(x)$ has a local maximum value at the boundary point $x = -1$ and a local minimum

value at the boundary point $x = 5$.

7. Let $f(x) = x(x+3)^{3/2} + 5x^2$ for $x \geq -3$. Show that $f(x)$ has a local maximum value at the boundary point $x = -3$.

8. Let $f(x) = \sqrt{x-2} + 5x$ for $x \geq 2$. Does the function $f(x)$ have a local maximum value or a local minimum value at the boundary point $x = 2$?

9. Suppose $f(x)$ is a continuous function only defined for $0 \leq x \leq 6$ that has all the following properties.

a) $f'(2) = DNE$ and $f'(4) = 0$

b) $f'(x) > 0$ for $0 < x < 2$ and for $4 < x < 6$.

c) $f'(x) < 0$ for $2 < x < 4$.

Find all the values of x for which this function has a local maximum. Find all the values of x for which this function has a local minimum.

6362 Finding Local Maxima and Local Minima

Suppose we have a function defined for $x \geq a$ only or for $x \leq b$ only or for $a \leq x \leq b$ only. How do we find the values of x which produce a local maximum value of x or a local minimum value of x? If we are looking at a value of x, say $x = c$, that is not an end point of the interval of definition, $(c \neq a, c \neq b)$, then the rules about finding local maximum and local minimum values are the same rules that we have already discussed. For end points we modified our definition of local maximum and local minimum to say what we mean by a one sided local maximum or a one sided local minimum. At the points $x = a$ and $x = b$ there can only be a one sided local maximum or a one sided local minimum. We have developed tests to determine if $x = a$ and $x = b$ are values of x that give a one sided local maximum or one sided local minimum value of $f(x)$. The most important fact when discussing end points is that it is very difficult to find a function $g(x)$ defined on the interval $a \leq x \leq b$ for which $x = a$ and $x = b$ do not give either a one sided local maximum or a one sided local minimum value.

Example 1. Let $f(x) = (x-4)^{4/3} - 4x$. Find all the values of x such that $f(x)$ has a local maximum value or a local minimum value at these numbers.

Solution. Note that $f(x)$ is defined for all x. There are no end points.

$$f'(x) = (4/3)(x-4)^{1/3} - 4.$$

The derivative $f'(x)$ is defined for all x. For what values of x is $f'(x) = 0$?

$$(4/3)(x-4)^{1/3} - 4 = 0$$
$$(x-4)^{1/3} = 3$$
$$(x-4) = 27$$
$$x = 31$$

We have $f'(31) = 0$. Thus $x = 31$ is the only critical value. We can test this critical value using either the First Derivative Test or the Second Derivative Test. Since the Second Derivative Test is easier to apply we begin by trying the Second Derivative Test.

$$f''(x) = (4/9)(x-4)^{-2/3}$$

This is defined for all x except $x = 4$. We have $f''(31) = (4/9)(31-4)^{-2/3} = 4/81$. Thus $f''(31) > 0$. Applying the Second Derivative Test we conclude that $f(x)$ has a local minimum value at $x = 31$. This minimum value is $f(31) = -43$.

We do not need to but let us determine the sign of the first derivative. If $x < 31$, then $x - 4 < 27$ and $(x - 4)^{1/3} < 3$ and $(4/3)(x - 4)^{1/3} < 4$ and $(4/3)(x - 4)^{1/3} - 4 < 0$. If $x < 31$, then $f'(x) < 0$. On the other hand, if $x > 31$, then $x - 4 > 27$, $(x - 4)^{1/3} > 3$, $(4/3)(x - 4)^{1/3} > 4$, and $(4/3)(x - 4)^{1/3} - 4 > 0$. If $x > 31$, then $f'(x) > 0$. Using this we could apply the First Derivative Test and conclude that $f(x)$ has a local minimum value at $x = 31$.

Example 2. Let $f(x) = 8(x + 3)^{3/2} - 3x^2$ for $x \geq -3$. Find all the values of x such that $f(x)$ has either a local maximum value or a local minimum value.

Solution. Note that $\sqrt{x + 3}$ is not defined for $x < -3$. We have $x = -3$ is an end point (boundary point) and so must be tested for local maximum or local minimum value.

$$f'(x) = 12(x + 3)^{1/2} - 6x \text{ for } x > -3.$$

Note that $f'(x)$ is found using the usual rules for finding the derivative. For what values of x is $f'(x) = 0$?

$$12(x + 3)^{1/2} - 6x = 0$$
$$(x + 3)^{1/2} = x/2$$
$$(x + 3) = (1/4)x^2$$
$$x^2 - 4x - 12 = 0$$
$$(x - 6)(x + 2) = 0$$
$$x = 6 \text{ and } x = -2.$$

We have $f'(6) = 12(9)^{1/2} - 6(6) = 0$, but $f'(-2) = 12(1)^{1/2} + 12 = 24$. The value $x = -2$ is an extraneous root. The only critical value is $x = 6$. We apply the Second Derivative Test to the critical value $x = 6$.

$$f''(x) = 6(x + 3)^{-1/2} - 6$$
$$f''(6) = 6(6 + 3)^{-1/2} - 6 = -4.$$

185

Applying the Second Derivative Test we conclude that $f(x)$ has a local maximum value at $x = 6$. Note that $f(6) = 108$.

Since the formula $f'(x) = 12(x + 3)^{1/2} - 6x$ is correct for $x > -3$ and the expression $12(x + 3)^{1/2} - 6x$ is continuous at $x = -3$, we can find the value of $f'(-3)$ by substituting into this formula. Substitution gives

$$f'(-3) = 18.$$

The value $x = -3$ is a left end point. Since $f'(-3) > 0$, we apply the Left End Point Theorem and conclude that $f(x)$ has a local minimum value at $x = -3$. Note that $f(-3) = -27$.

Example 3. Let $f(x) = 2|x - 1| - |x - 5|$. For what values of x does $f(x)$ have a local maximum value and for what values does $f(x)$ have a local minimum value. Note that $f(x)$ is continuous.

Solution. Note that $f(x)$ is defined for all values of x. There are no boundary points. It is easier to analyze functions of this kind if they are first written as a function in parts. If $x < 1$, then $|x - 1| = -(x - 1)$ and $|x-5| = -(x-5)$. If $x < 1$, then $2|x-1|-|x-5| = -2(x-1)+|x-5| = -x-3$. If $1 < x < 5$, then $|x - 1| = x - 1$ and $|x - 5| = -(x - 5)$. If $1 < x < 5$, then $2|x - 1| - |x - 5| = 2(x - 1) + (x - 5) = 3x - 7$. If $x > 5$, then $|x - 1| = x - 1$ and $|x-5| = x-5$. If $x > 5$, then $2|x-1|-|x-5| = 2(x-1)-(x-5) = x+3$. Written as a function in parts $f(x)$ is

$$f(x) = \begin{cases} -x - 3 & x \leq 1 \\ 3x - 7 & 1 \leq x \leq 5 \\ x + 3 & x \geq 5. \end{cases}$$

We can now easily compute the derivative

$$f'(x) = \begin{cases} -1 & x < 1 \\ 3 & 1 < x < 5 \\ 1 & x > 5. \end{cases}$$

Note that $\lim_{x \to 1-} f'(x) = -1$ and $\lim_{x \to 1+} f'(x) = 3$ and so $\lim_{x \to 1} f'(x) = DNE$. Also $\lim_{x \to 5-} f'(x) = 3$ and $\lim_{x \to 5+} f'(x) = 1$ and so $\lim_{x \to 5} f'(x) = DNE$, we conclude that $f'(1) = DNE$ and $f'(5) = DNE$. Since $f'(x) = 0$ is not true for

any x, the critical points are $x = 1$ and $x = 5$. For the record $f''(1) = DNE$ and $f''(5) = DNE$. Also $f''(x) = 0$ for all other values of x. We can not apply the Second Derivative Test. The sign of the first derivative is

$$\underset{\underset{\displaystyle 1}{}}{f'(x) < 0} \quad \bigg| \quad \underset{\underset{\displaystyle 5}{}}{f'(x) > 0} \quad \bigg| \quad f'(x) > 0 \qquad\qquad x$$

Applying the First Derivative Test we conclude that $f(x)$ has a local minimum value at $x = 1$. Since $f'(x) > 0$ for both $1 < x < 5$ and $x > 5$, we conclude that $f(x)$ is increasing at $x = 5$. The function $f(x)$ has neither a local maximum value nor a local minimum value at the critical point $x = 5$.

Example 4. Let $f(x) = 4(x - 1)^{3/2} + 2(6 - x)^{3/2}$ for $1 \leq x \leq 6$. Find all the values of x such that $f(x)$ has either a local maximum value or a local minimum value.

Solution. Note that $\sqrt{x - 1}$ is not defined for $x < 1$ and $\sqrt{6 - x}$ is not defined for $x > 6$. The values $x = 1$ and $x = 6$ are end points (boundary points) and so must be tested.

$$f'(x) = 6(x - 1)^{1/2} - 3(6 - x)^{1/2} \text{ for } 1 < x < 6.$$

We found $f'(x)$ using the usual rules when $1 < x < 6$. Since this expression for $f'(x)$ is continuous at $x = 1$ and $x = 6$, we can find $f'(1)$ and $f'(6)$ by substituting into the expression for $f'(x)$ for $1 < x < 6$.

$$f'(1) = -3\sqrt{5}$$
$$f'(6) = 6\sqrt{5}.$$

We need to find the critical points where $f'(x) = 0$.

$$6(x - 1)^{1/2} - 3(6 - x)^{1/2} = 0$$
$$2(x - 1)^{1/2} = (6 - x)^{1/2}$$
$$4(x - 1) = 6 - x$$
$$5x = 10$$
$$x = 2.$$

Substituting we get $f'(2) = 6\sqrt{1} - 3\sqrt{4} = 0$. The critical value is $x = 2$. Note that $x = 2$ is not an end point of the domain of $f(x)$. We will try the Second Derivative Test first.

$$f''(x) = 3(x-1)^{-1/2} + (3/2)(6-x)^{-1/2}$$
$$f''(2) = 3(2-1)^{-1/2} + (3/2)(6-2)^{-1/2} = 15/4.$$

Thus $f''(2) > 0$. Applying the Second Derivative Test we conclude that $f(x)$ has a local minimum value at $x = 2$. Note $f(2) = 20$. Next check the end points. Since $f'(1) = -3\sqrt{5} < 0$ we apply the Left End Point Theorem at a left end point and conclude that $f(x)$ has a local maximum value at $x = 1$. Note that $f(1) = 2(5)^{3/2} = 10\sqrt{5}$. Since $f'(6) = 6\sqrt{5} > 0$ we apply the Right End Point Theorem and conclude that $f(x)$ has a local maximum value at $x = 6$. Note that $f(6) = 20\sqrt{5}$.

Just for fun let us see how much work it is to apply the First Derivative Test. In order to apply the First Derivative Test at the critical point $x = 2$ we would first note that if $x < 2$, $x - 1 < 1$, $\sqrt{x-1} < 1$ and $6\sqrt{x-1} < 6$. Also if $x < 2$, $-x > -2$, $6 - x > 4$, $\sqrt{6-x} > 2$, and $-3\sqrt{6-x} < -6$. It follows that $6\sqrt{x-1} - 3\sqrt{6-x} < 0$. In a similar way if $x > 2$, then $6\sqrt{x-1} - 3\sqrt{6-x} > 0$. Therefore, the first derivative goes from negative to positive at $x = 2$. This is more work than finding the second derivative.

Example 5. Let $f(x) = [4+3x-x^2]^{1/2}$ for $-1 \leq x \leq 4$. Find all the values of x such that $f(x)$ has either a local maximum value or a local minimum value.

Solution. The end points are $x = -1$ and $x = 4$. Using the chain rule we find

$$f'(x) = \frac{3 - 2x}{2\sqrt{4 + 3x - x^2}} \quad \text{for } -1 < x < 4.$$

In order for a fraction to equal zero the numerator must be zero. We have $3 - 2x = 0$ when $x = 3/2$. Thus $f'(3/2) = 0$ and $x = 3/2$ is a critical point. We could test this critical point using the Second Derivative Test which requires we find $f''(x)$. For this problem it is easier to use the First Derivative Test. We first note that $4 + 3x - x^2 > 0$ for $-1 < x < 4$. This means that $2\sqrt{4 + 3x - x^2} > 0$, or positive. Clearly, $3 - 2x > 0$ if $x < 3/2$ and $3 - 2x < 0$ if $x > 3/2$. It follows that $f'(x) > 0$ for $-1 < x < 3/2$ and

$f'(x) < 0$ for $3/2 < x < 4$. Applying the First Derivative Test we conclude that $f(x)$ has a local maximum value at $x = 3/2$.

Now let us look at the end points $x = -1$ and $x = 4$. Note that $\lim_{x \to -1+} f'(x) = +\infty$ and $\lim_{x \to 4-} f'(x) = -\infty$. This means that $f'(-1) = DNE$ and $f'(4) = DNE$. Applying the Left End Point Theorem we conclude that $f(x)$ has a local minimum value at $x = -1$. Applying the Right End Point Theorem we conclude that $f(x)$ has a local minimum value at $x = 4$. Note that $f(1) = 0$ and $f(4) = 0$.

Example 6. Let $f(x)$ denote a continuous function defined on the closed interval $-2 \leq x \leq 6$ with the following properties. Determine the critical points and end points for $f(x)$. Determine if $f(x)$ has a local maximum value or a local minimum value at these points.

a) $f'(0) = DNE$, $f'(2) = 0$, $f'(4) = DNE$, $f''(2) = 0$, $f'(-2) > 0$ and $f'(6) > 0$. Note that $f'(-2)$ and $f'(6)$ exist.

b) $f'(x) > 0$ for $-2 \leq x \leq 0$ and for $4 < x \leq 6$. Also $f'(x) < 0$ for $0 < x < 2$ and for $2 < x < 4$.

c) $f''(x) > 0$ for $-2 < x < 0$, for $0 < x < 2$, and for $4 < x < 6$. Also $f''(x) < 0$ for $2 < x < 4$.

Solution. Since $f'(0) = DNE$ and $f'(4) = DNE$ it must also be true that $f''(0) = DNE$ and $f''(4) = DNE$. The critical points for this function are $x = 0$, $x = 2$, and $x = 4$. These are all interior points.

	$f''(x) > 0$ $f'(x) > 0$		$f''(x) > 0$ $f'(x) < 0$		$f''(x) < 0$ $f'(x) < 0$		$f''(x) > 0$ $f'(x) > 0$	
-2		0		2		4		6

Consider $x = 0$. For $-2 < x < 0$ we have $f'(x) > 0$ and for $0 < x < 2$, we have $f'(x) < 0$. Applying the First Derivative Test we conclude that $f(x)$ has a local maximum value at $x = 0$.

Consider $x = 2$. For $0 < x < 2$ we have $f'(x) < 0$ and for $2 < x < 4$ we have $f'(x) < 0$. It follows that $f(x)$ is decreasing for $x = 2$. The function $f(x)$ has neither a local maximum value nor a local minimum value at $x = 2$.

Consider $x = 4$. For $2 < x < 4$ we have $f'(x) < 0$ and for $4 < x < 6$ we have $f'(x) > 0$. Applying the First Derivative Test we conclude that $f(x)$ has a local minimum value at $x = 4$.

Since $f''(0) = DNE$, $f''(2) = 0$, and $f''(4) = DNE$ no information about maximum and minimum values can be obtained from the Second Derivative Test.

Note that $f'(-2) > 0$. Applying the Left End Point Theorem we conclude that $f(x)$ has a local minimum value at $x = -2$.

Note that $f'(6) > 0$. Applying the Right End Point Theorem we conclude that $f(x)$ has a local maximum value at $x = 6$.

Example 7. Consider the function

$$f(x) = \frac{x^2}{x^2 - 9}$$

Does this function have any local maximum values or local minimum values? This function is defined for all values of x except $x = -3$ and $x = 3$. We really should write

$$f(x) = \frac{x^2}{x^2 - 9} \quad \text{for } x \neq -3 \text{ and } x \neq 3.$$

Since $f(-3)$ is not defined the question "does the function $f(x)$ have a local maximum value at $x = -3$?" does not make any sense. In particular, we do not say that $f(-3)$ equals infinity. We can look at all the values of x for which $f(x)$ is defined to see if these values give a local maximum or a local minimum. This function is not defined for $x = -3$ and $x = 3$. These are not end points for the domain of definition.

In order to find the critical values of x we need the first derivative.

$$f'(x) = \frac{-18x}{(x^2 - 9)^2}.$$

We see that $f'(0) = 0$. This means that $x = 0$ is a critical value. Also $f'(-3)$ and $f'(3)$ do not exist. But the function $f(x)$ does not exist when $x = -3$ and $x = 3$. This means that $x = -3$ and $x = 3$ can not be critical values of $f(x)$. Note that if $-3 < x < 0$, then $f'(x) > 0$. Also if $0 < x < 3$, then $f'(x) < 0$. The First Derivative Test tells us that $f(x)$ has a local maximum value at $x = 0$. There are no local minimum values of $f(x)$.

Example 8. Consider the function $f(x) = x^{-2}$. Does this function have any local maximum or local minimum values?

Solution. First, we clearly see that $f(0)$ is not defined. Also, $f'(x) = -2x^{-3}$ for $x \neq 0$. The value $f'(0)$ does not exist. Since $f'(0) = DNE$ we might think $x = 0$ is a critical point. We do not say $x = 0$ is a critical point. The reason is that $f(0)$ does not exist. Note that if $x < 0$, then $f'(x) > 0$ and if $x > 0$, then $f'(x) < 0$. Hence, if $x < 0$, then $f(x)$ is increasing. If $x > 0$, then $f(x)$ is decreasing. This makes it clear that $f(x)$ does not have a local maximum value or a local minimum value. However, it is interesting to look at some limits.

$$\lim_{x \to 0+} \frac{1}{x^2} = +\infty \qquad \lim_{x \to 0-} \frac{1}{x^2} = +\infty$$

$$\lim_{x \to +\infty} \frac{1}{x^2} = 0 \qquad \lim_{x \to -\infty} \frac{1}{x^2} = 0.$$

Exercises

1. Let $f(x) = 2(2x + 24)^{3/2} - 3x^2$ for $x \geq -12$. Find all the values of x such that $f(x)$ has either a local maximum value or a local minimum value.

2. Let $f(x) = \begin{cases} -x - 3 & x < 0 \\ -x^2 + 4x - 3 & 0 \leq x \leq 3 \\ 2x - 6 & x > 3. \end{cases}$
Find all the values of x such that $f(x)$ has either a local maximum value or a local minimum value. Note that $f(x)$ is continuous for all values of x. Hint: use the first derivative test.

3. Let $f(x) = 6(x - 3)^{3/2} + 2(13 - x)^{3/2}$ for $3 \leq x \leq 13$. Find all the values of x such that $f(x)$ has either a local maximum value or $f(x)$ has a local minimum value.

4. Let $f(x) = [5 + 4x - x^2]^{3/4}$ for $-1 \leq x \leq 5$. Find all the values of x such that $f(x)$ has either a local maximum value or $f(x)$ has a local minimum value.

5. Let $f(x) = 8x^{1/3} - x^{4/3}$ for all x. Find all the values of x such that $f(x)$ has either a local maximum value or $f(x)$ has a local minimum value.

6. Let $f(x) = \dfrac{x^2}{16 - x^2}$ for all x. Are there any values of x which give either a local maximum value or a local minimum value of $f(x)$?

7. Let $f(x) = 2(x - 5)^{1/2} + (50 - x)^{1/2}$ for $5 \le x \le 50$. Find all the values of x such that $f(x)$ has a local maximum value or a local minimum value.

8. Let $f(x) = (25 - x^2)^{3/2} + 6x^2$ for $-5 \le x \le 5$. Show that $x = -3$, $x = 0$, $x = 3$ are critical points. Classify these critical points and the end points as points where $f(x)$ has a local maximum value, a local minimum value or neither.

9. The function $f(x)$ is a continuous function defined for $x \ge 0$ with all the following properties:

a) $f'(0) = 2$, $f'(2) = 0$, $f'(4) = DNE$, $f'(6) = 0$, $f''(2) = 0$, $f''(4) = DNE$, $f''(6) \ne 0$.

b) $f'(x) > 0$ for $0 < x < 2$, for $2 < x < 4$, and for $x > 6$. Also $f'(x) < 0$ for $4 < x < 6$.

c) $f''(x) > 0$ for $2 < x < 4$, for $4 < x < 6$, and for $x > 6$. Also $f''(x) < 0$ for $0 < x < 2$.

Find all the critical points of $f(x)$ classify these critical points and the end points as points where $f(x)$ has a local maximum value, a local minimum value, or neither.

6365 Optimization

The methods we have learned for finding extreme values can be applied to solving many practical problems. For example, these methods enable us to minimize cost or to minimize time. These problems are called optimization problems. Optimization problems occur in a wide range of applications. The greatest challenge in solving an optimization problem is to set up the function that is to be maximized or minimized.

Example 1. A farmer has 3400 feet of fencing and wants to fence off a field in the shape of a rectangle. He wants to divide the field into 3 pens using the fence. What are the dimensions of the field that has the largest area?

Solution. We need to set up an equation which connects the variables in the problem. The first step is to identify the variables. In this problem the quantities that change are length, width, and the area of the field. Let

$$L = \text{ length of the field}$$
$$w = \text{ length of the four short sides}$$
$$A = \text{ area of the field.}$$

$$A = Lw.$$

Which are the independent variables and which variable is the dependent variable? We look at the question asked by the problem. We are asked to maximize area. We think of area, A, as depending on length, L, and width, w. Area, A, is the dependent variable while L and w are independent variables. This is one variable calculus. When finding the maximum and minimum we can not have two independent variables. We must find a relationship between L and w. Look at a diagram of the fence.

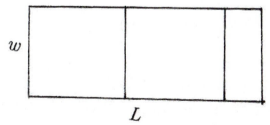

The total length of the fence is $4w + 2L = 3400$ so that $L = 1700 - 2w$.

Therefore,
$$A(w) = w(1700 - 2w) = 1700w - 2w^2.$$

The equation $A(w) = 1700w - 2w^2$ expresses area as a function of width. The width, w, is the independent variable. What are the possible values of w? Clearly, $0 \leq w \leq 850$. In fact $w = 0$ is really not possible. This is a practical problem. There is no enclosure if $w = 0$ and $L = 850$. Also $w = 850(L = 0)$ is not possible. If we assume that these are possible values of w, we easily get that $A(0) = 0$ and $A(850) = 0$. Therefore, let us take the position that

$$A(w) = 1700w - 2w^2 \text{ and } 0 < w < 850,$$

describes the relationship between w and A. If $w = b$ with $0 < b < 850$, then $A(w)$ is defined for $w < b$ and w close to b. Also $A(w)$ is defined for $w > b$ and w close to b. The point b is an interior point. There are no end points (boundary points). In order to find critical points we find $A'(w)$.

$$A'(w) = 1700 - 4w$$
$$A'(425) = 0$$

The critical point is $w = 425$. The second derivative test is easy

$$A''(w) = -4.$$

Applying the second derivative test we conclude that $A(w)$ has a local maximum value at $w = 425$. Since this is the only local maximum it must be the absolute maximum.

$$A(425) = 425[1700 - 2(425)] = 361,250.$$

We have $w(1700 - 2w) \leq 361,250$ for $0 < w < 850$. When $w = 425$, $L = 1700 - 2(425) = 850$. The maximum area is 361,250 ft^2.

Example 2. Suppose we are required to construct a rectangular room with a square floor. The cost of material for the top of the room is \$12 per square foot. The cost of material for the floor of the room is \$18 per ft^2. The cost of material for the four sides is \$21 per ft^2. The volume of the space inside the room must be 1960 ft^3. Find the dimensions of the room that minimize the cost of material.

Solution. First, we assign a letter to each variable. Let

$$x = \text{ the length of one side of the square floor}$$
$$y = \text{ height of the room}$$
$$c = \text{ total cost of material to build the room.}$$

We are asked to minimize cost. We must find an expression for the cost involving x and y. The cost of the floor is $12x^2$ and the cost of the top is $18x^2$. The cost of each of the four walls is $21xy$. The total cost of materials is

$$c = 12x^2 + 18x^2 + 4(21xy).$$

We need an equation connecting the variables x and y. The volume of the room is x^2y. We are given that the volume must be 1960. Thus, $x^2y = 1960$ or $y = 1960x^{-2}$. Substituting this into the cost function, we get

$$c(x) = 30x^2 + 84x(1960x^{-2}) = 30x^2 + 164640x^{-1}.$$

Since x and y are both lengths it is clearly true that $x > 0$ and $y > 0$. Since $x^2y = 1960$ it is also true that x and y can be any numbers greater than zero. The number y can be very large if x is very near zero. We are looking for a local minimum value for the function $c(x)$ for $x > 0$. There are no end points.

$$c'(x) = 60x - 164640x^{-2}.$$

In order to find the critical points we set $c'(x) = 0$.

$$60x - 164640x^{-2} = 0$$
$$60x^3 = 164640$$
$$x^3 = 2744$$
$$x = 14.$$

The second derivative test is easy.

$$c''(x) = 60 + 2(164640)x^{-3}$$
$$c''(14) = 60 + 120 = 180 > 0.$$

The second derivative test tells us that the function $c(x)$ has a local minimum value at $x = 14$. There is only one local minimum value.

$$x^2 y = 1960 \qquad (14)^2 y = 1960 \qquad y = 10$$

The minimum value for the cost is

$$c(14) = 30(14)^2 + 84(14)(10) = \$17,640.$$

In order to apply the first derivative test we would need to write the first derivative as $c'(x) = 60x^{-2}(x - 14)(x^2 + 14x + 196)$.

Example 3. A certain farmer has a field of potatoes. On August 1 there is an average of 560 bushels of potatoes per acre. For each day that the farmer leaves the potatoes in the field the size of the crop increases by an average of 40 bushels per acre. On August 1 the price of potatoes is \$24 per bushel. However, for every day after August 1 the price of potatoes decreases by \$0.60 per bushel. In order to get the maximum amount of money how many days after August 1 should the farmer leave the potatoes in the field before selling them.

Solution. Let t equal the number of days that the potatoes are left in the field after August 1. Let $c(t)$ denote the amount of money that the potatoes are worth if they are left in the field t days after August 1. The average number of bushels of potatoes per acre after August 1 is

$$560 + 40t$$

The price for a bushel of potatoes after t days is

$$24 - (0.6)t.$$

The total selling price for an acre of potatoes after t days is

$$c(t) = (560 + 40t)(24 - 0.6t).$$

In order to find the critical numbers for the function $c(t)$ we find the derivative $c'(t)$.

$$c'(t) = (560 + 40t)(-0.6) + (24 - 0.6t)(40)$$
$$= 624 - 48t.$$

Setting $c'(t) = 0$, we get $48t = 624$ or $t = 13$. The critical number is $t = 13$.

$$c''(t) = -48.$$

By the Second Derivative Test the critical point $t = 13$ is a point where $c(t)$ has a local maximum. In order for the potatoes to be worth the most money the farmer should let them grow another 13 days.

Example 4. A river has sides that are straight lines and is 441 yards wide. A powerline must be built from point A on one side of the river to point B on the other side. Point A is 2,000 yards down river from point D which is directly across the river from point B. The cost of building the powerline is $800 per yard over land and $1,000 per yard over water. How many yards down river from point D should the line enter the water in order to have the lowest cost?

Solution. It is probably best to start with a diagram.

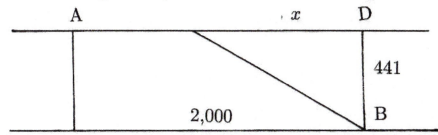

Let x denote the distance from the point where the powerline enters the water to point D. The distance from the point where the powerline enters the water to point A is $2000 - x$. The length of the powerline over the water is

$$\sqrt{x^2 + (441)^2}$$

The total cost of the powerline is

$$c(x) = 800(2000 - x) + 1000(x^2 + 441^2)^{1/2}$$

We are making the logical assumption that the problem is set up as shown with $0 \le x \le 2000$. However, suppose that the power line is being built by an insane person who thinks that the power line should go past point D before entering the water or who thinks it should start at a point by first going away from point A. Either of these assumptions would require a different cost function. In that case, we would need to consider that x can be any real number.

We need to find the critical points of the cost function $c(x)$.

$$c'(x) = -800 + 1000(1/2)(x^2 + 441^2)^{-1/2}(2x)$$
$$= -800 + \frac{1000x}{\sqrt{x^2 + 441^2}}$$

Setting $c'(x) = 0$, we get

$$\frac{x}{\sqrt{x^2 + 441^2}} = \frac{4}{5}.$$

Squaring both sides of the equation, we have

$$x^2 + 441^2 = (25/16)x^2$$
$$441^2 = (9/16)x^2$$
$$x = \frac{4}{3}(441) = 588 \text{ yards}$$

Note that $c'(-588) = -1600 \neq 0$. We can use the Second Derivative Test to determine if $c(x)$ has a local maximum or a local minimum at $x = 588$.

$$c''(x) = \frac{1000(441^2)}{(x^2 + 441^2)^{3/2}}$$
$$c''(588) = \frac{1000(441^2)}{(588^2 + 441^2)^{3/2}} = \frac{1000(441^2)}{735^3} > 0.$$

Since $c''(588) > 0$ we conclude applying the Second Derivative Test that $c(x)$ has a local minimum at $x = 588$.

$$c(588) = 800[2000 - (588)] + 1000[(588)^2 + (441)^2]^{1/2}$$
$$= \$1,864,600$$

Since $c(x)$ is defined on the closed interval $0 \leq x \leq 2000$, then we must test the end points $x = 0$ and $x = 2000$ to see if $f(x)$ has a local maximum or a local minimum for these values of x.

$$c'(x) = -800 + \frac{1000x}{\sqrt{x^2 + 441^2}}$$

$$c'(0) = -800$$

$$c'(2000) = -800 + \frac{1000(2000)}{\sqrt{2000^2 + 441^2}} = 176.54.$$

Since $c'(0) < 0$, then applying the Left End Point Theorem we conclude that $c(x)$ has a local maximum value at $x = 0$. Since $c'(2000) > 0$, then applying the Right End Point Theorem we conclude that $c(x)$ has a local maximum value at $x = 2000$.

$$c(0) = 800(2000) + 1000(441) = 2,041,000$$

$$c(2000) = 1000(2000^2 + 441^2)^{1/2} = 2,048,043$$

The cost function $c(x)$ has an minimum value at $x = 588$ of $c(588) = \$1,864,600$.

Exercises

1. A farmer has 4300 feet of fencing and wants to fence off a field in the shape of a rectangle. He wants to divide the field into 4 pens using the fence. What are the dimensions of the field that has the largest area?

2. A farmer has 1160 yards of fencing and wants to fence off a field in the shape of a rectangle. He wants to divide the field into 3 pens. The field borders a river and so he needs no fence along the river. What are the dimensions of the field that has the largest area?

3. A box with an open top is to be constructed with a volume of 2304 in^3. The length of the base must be twice the width of the base. Find the dimensions of the box that has the smallest surface area.

4. If 2028 cm^2 of material is available to make a box with a square base and an open top, find the largest possible volume of the box.

5. Suppose we are required to construct a rectangular room with a square floor. The cost of material for the top of the room is $13 per square foot.

The cost of material for the floor of the room is $20 per ft^2. The cost of material for the four sides is $22 per ft^2. The volume of space inside the room must be 6,000 ft^3. Find the dimensions of the room that minimize the cost of material.

6. Suppose we must construct a rectangular tank such that the length of the base is 2 times the width of the base and which has no top. The cost of material for the bottom of the tank is $21 per ft^2. The cost of the material for the four sides of the tank is $49 per ft^2. The volume of the space inside the tank must be 1568 ft^3. Find the dimensions of the tank that minimizes the cost of material.

7. The owner of an apple orchard estimates that if 24 trees are planted per acre, then each mature tree will yield 600 apples per year. For each additional tree planted per acre the number of apples produced by each tree decreases by 12 per year. How many trees should be planted per acre to obtain the most apples per year?

8. A certain oil field contains 10 wells which produce 240 barrels of oil each per day. For each additional well that is drilled, the average production per well decreases by 12 barrels per day. How many additional wells should be drilled to obtain the maximum amount of oil per day?

9. Suppose we must construct a container in the form of a cylinder with a bottom but without a top. The cost of material for the side is $10 per ft^2. The cost of material for the bottom is $15 per ft^2. The volume of the container must be 2592π ft^3. What are the dimensions of the cylinder that minimizes the cost of material?

10. The strength of a rectangular beam is directly proportional to the product of its width w and the square of the depth, s, of a cross section. Find the dimensions of the strongest beam that can be cut from a cylindrical log of radius r.

11. A river has sides that are straight lines and is 300 yards wide. A powerline must be built from point A on one side of the river to point B on the other side. Point A is 4,000 yards down river from point D which is directly across the river from point B. The cost of building the powerline is $720 per yard over land and $900 per yard over water. How many yards down river from point D should the line enter the water in order to have the lowest cost?

12. Mary is 2 miles off shore in a boat and wishes to reach a point on the coast 6 miles down a straight shore line from the point nearest the boat. She can row 3mph and walk 5mph. Where should she land the boat in order to reach the point on the shore in the least amount of time. Hint: You need to recall the formula distance equals rate times time.

13. A piece of wire 20 meters long is cut into two pieces. One piece is bent into a square and the other is bent into a circle. How should the wire be cut so that the enclosed area is a maximum.

14. A cable television firm presently serves 5000 households and charges $20 per month. A marketing survey indicates that each decrease of $1 in monthly charge will result in 500 new customers. Find the monthly charge that will result in the maximum monthly revenue.

6367 Review of Graphing

One of the best ways to understand what is happening to the values of a function $f(x)$ as x increases is to look at a graph of the function. Given the graph of $y = f(x)$ we can reach lots of conclusions about the function $f(x)$ and the first two derivatives. In order to become more familiar with the relationship between the function $f(x)$ and the graph of $y = f(x)$ we will practice starting with some conditions on $f(x)$, on $f'(x)$, and $f''(x)$ and using only these conditions sketch the graph of $y = f(x)$. We will sketch such graphs only in the case that the function $f(x)$ is continuous. Sketching graphs for functions $f(x)$ such that $f(x)$ is not continuous would require us to develop the definitions of several terms which have so far not been developed.

Example 1. Sketch a graph for a continuous function $f(x)$ defined for all x that satisfies all the following conditions.

(a) $f'(-3) = 0$, $f'(1) = DNE$, $f'(4) = DNE$, $f''(-3) > 0$, $f''(1) = DNE$, and $f''(4) = DNE$.

(b) $f'(x) > 0$ for $-3 < x < 1$ and for $x > 4$. Also $f'(x) < 0$ for $x < -3$ and for $1 < x < 4$.

(c) $f''(x) > 0$ for $x < -3$ and for $-3 < x < 1$. Also $f''(x) < 0$ for $1 < x < 4$ and for $x > 4$.

Solution. We first draw a number line with the conditions on the derivatives on each interval indicated.

$f' < 0$ $f'' > 0$	$f' > 0$ $f'' > 0$	$f' < 0$ $f'' < 0$	$f' > 0$ $f'' < 0$
-5 $\quad$ -3	1	4	6

No specific values of $f(x)$ are given. We sketch a graph such that $f'(x) < 0$ and $f''(x) > 0$ for $x < -3$. Next, on the interval $-3 < x < 1$ we sketch the graph such that $f'(x) > 0$ and $f''(x) > 0$. We continue in the same manner on the other intervals being careful to keep the graph continuous.

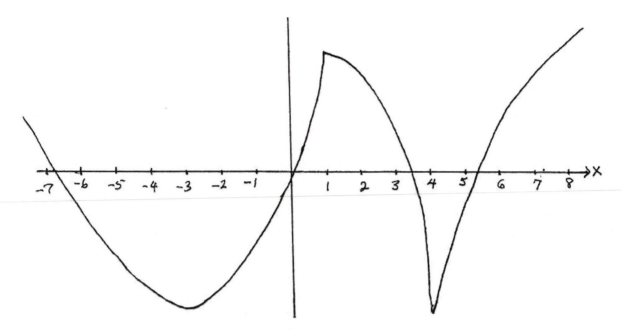

Suppose we are given a function $f(x)$ and wish to determine for what values of x the function $f(x)$ has a local maximum and for what values of x it has a local minimum. The solution of this type of problem is done in two steps. The first step is starting with the function determine its critical points, determine where the function is increasing and where it is decreasing, and where the function is concave up and where it is concave down. The second step is to use this information along with either the First Derivative Test or the Second Derivative Test to determine for which critical points the function has a local maximum value or a local minimum value.

Example 2. Sketch the graph of a continuous function defined for all real numbers that satisfies all the following properties. Sketch the section of the graph from $x = -4$ to $x = 6$.

(a) $f'(-2) = DNE$, $f'(1) = 0$, $f'(4) = DNE$, $f''(-2) = DNE$, $f''(1) = 0$, and $f''(4) = DNE$.

(b) $f'(x) > 0$ for $-2 < x < 1$ and for $1 < x < 4$. Also $f'(x) < 0$ for $x < -2$ and for $x > 4$.

(c) $f''(x) > 0$ for $1 < x < 4$. Also $f''(x) < 0$ for $x < -2$, for $-2 < x < 1$, and for $x > 4$.

Solution. Note that there are no exact values of $f(x)$ given. It is usually a good idea when sketching a curve to make the following chart.

$$f''(x) < 0 \qquad f''(x) < 0 \qquad f''(x) > 0 \qquad f''(x) < 0$$
$$f'(x) < 0 \qquad f'(x) > 0 \qquad f'(x) > 0 \qquad f'(x) < 0$$

$$\overbrace{}$$
$$\underset{-2}{|} \qquad\qquad \underset{1}{|} \qquad\qquad \underset{4}{|}$$

We begin by sketching a curve for $x < -2$ such that $f'(x) < 0$ and $f''(x) < 0$, that is, a curve that is decreasing and concave down. Next, for $-2 < x < 1$ we sketch a curve that is increasing and concave up.

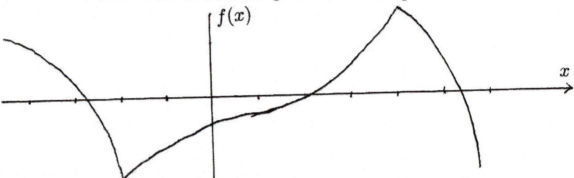

Once we have constructed this graph we can see which values of x give a maximum value and which values of x give a minimum value for the function $f(x)$. We can also see for what values of x the function is increasing and for what values of x the function is decreasing. However, if we are given the problem to find the values of x for which the function has a minimum value, the values of x for which the function is increasing, and so forth, then we must rely on the theorems. For example, we conclude that $x = -3$ gives a minimum value of $f(x)$ by noting that $f'(x) < 0$ for $x < -3$ and $f'(x) > 0$ for $x > -3$. We can then say that $f(x)$ has a local minimum value at $x = -3$ by the First Derivative Test. Alternately we could conclude that $f(x)$ has a local minimum value at $x = -3$ by applying the Second Derivative Test since $f''(3) > 0$. We conclude that $f(x)$ is decreasing for $1 < x < 4$ since $f'(x) < 0$ for $1 < x < 4$.

Example 3. The function $f(x)$ is a continuous function defined for all x with the following properties.

a) $f'(-2) = DNE, f'(1) = 0, \ f'(4) = DNE, f''(-2) = DNE, f''(1) > 0$ and $f''(4) = DNE$.

b) $f'(x) > 0$ for $x < -2$ and for $1 < x < 4$. Also $f'(x) < 0$ for $-2 < x < 1$ and for $x > 4$.

c) $f''(x) > 0$ for $-2 < x < 4$ and for $x > 4$. Also $f''(x) < 0$ for $x < -2$.

Find the critical points for $f(x)$. Use the First Derivative Test to determine which critical values give a local maximum value of $f(x)$ and which give a local minimum value of $f(x)$. Second, use the Second Derivative Test to determine which critical values give a local maximum value and which give a local minimum value of $f(x)$. Sketch a possible graph of $f(x)$.

Solution. The critical values of x for $f(x)$ are $x = -2$, $x = 1$, and $x = 4$. There are no end points for this function since it is defined for all values of x.

Let us test the critical values using the First Derivative Test.
When $x < -2$, then $f'(x) > 0$.
When $-2 < x < 1$, then $f'(x) < 0$.

Applying the First Derivative Test, we conclude that $f(x)$ has a local maximum value at $x = -2$. Since we have no formula for $f(x)$ we have no idea what the value of $f(-2)$ is.

Next apply the First Derivative Test to the critical value $x = 1$.

When $-2 < x < 1$, $f'(x) < 0$.
When $1 < x < 4$, $f'(x) > 0$.

The First Derivative Test says: if $f'(x) < 0$ when $x < 1$ and $|x - 1|$ is small and if $f'(x) > 0$ when $x > 1$ and $|x - 1|$ is small, then $f(x)$ has a local minimum value at $x = 1$. We conclude that $f(x)$ has a local minimum value at $x = 1$.

Apply the First Derivative Test to the critical value $x = 4$.

When $1 < x < 4$, $f'(x) > 0$.
When $x > 4$, $f'(x) < 0$.

Applying the First Derivative Test we conclude that $f(x)$ has a local maximum value at $x = 4$.

Next, we test the critical values using the Second Derivative Test. Since $f''(-2) = DNE$, the Second Derivative Test gives no information regarding whether the function $f(x)$ has a local maximum value or a local minimum value at $x = -2$.

Since $f'(1) = 0$ and $f''(-1) > 0$, we apply the Second Derivative Test and conclude that $f(x)$ has a local minimum value at $x = 1$.

Since $f''(4) = DNE$, the Second Derivative Test gives no information (fails) about whether the function $f(x)$ has a local maximum value or a local minimum value at $x = 4$.

Let us note that $x = -2$ is an inflection point of the graph of $f(x)$ even though $f''(-2) = DNE$. An inflection point is a point where the concavity of the graph of $f(x)$ changes. Note that even though $f''(4) = DNE$ there is no inflection point at $x = 4$.

Let us make the following observation. The First Derivative Test basically always works. The Second Derivative Test fails is some cases. The First Derivative Test is longer and harder to apply than the Second Derivative Test.

Next, let us sketch a graph of $f(x)$. First, make a chart of the facts about $f'(x)$ and $f''(x)$.

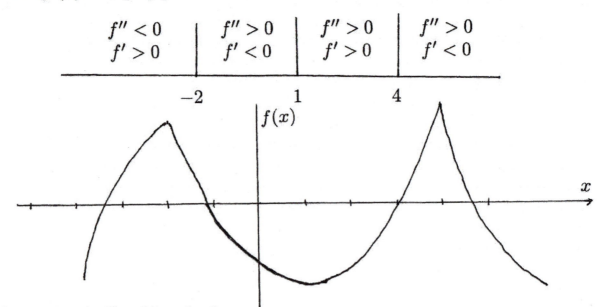

Example 4. Consider the function

$$f(x) = \frac{x^2 - x + 19}{x^2 + 16}$$

Find the values of x for which $f(x)$ has a local maximum and for which $f(x)$ has a local minimum.

Solution. Using the quotient rule to differentiate, we get

$$f'(x) = \frac{(x^2 + 16)(2x - 1) - (x^2 - x + 19)(2x)}{(x^2 + 16)^2}$$

This simplifies to

$$f'(x) = \frac{x^2 - 6x - 16}{(x^2 + 16)^2}.$$

First, note that $x^2 + 16 \geq 16 > 0$. The solutions of $x^2 - 6x - 16 = 0$ are $x = 8$ and $x = -2$. The critical points of $f(x)$ are $x = 8$ and $x = -2$, that is, $f'(8) = 0$ and $f'(-2) = 0$. Since $x^2 - 6x - 16 = (x - 8)(x + 2)$, we see that $x^2 - 6x - 16 > 0$ for $x < -2$ and for $x > 8$. Also $x^2 - 6x - 16 < 0$ for $-2 < x < 8$. It follows that

$$f'(x) > 0 \quad \text{for } x < -2 \text{ and for } x > 8,$$
$$f'(x) < 0 \quad \text{for } -2 < x < 8.$$

We can use the First Derivative Test to classify the critical points.

If $x < -2$, then $f'(x) > 0$. If $-2 < x < 8$, then $f'(x) < 0$.

It follows by the First Derivative Test that $f(x)$ has a local maximum value at $x = -2$. In fact $f(-2) = 5/4$.

If $-2 < x < 8$, then $f'(x) < 0$. If $x > 8$, then $f'(x) > 0$.

It follows by the First Derivative Test that $f(x)$ has a local minimum value at $x = 8$. Since $f(8) = 15/16$, this means $f(x) > (15/16)$ for $x \geq 8$. Since $f(-2) = 5/4$, this means $f(x) < (5/4)$ for $x < -2$. In fact, we can show using inequalities that $(15/16) \leq f(x) \leq 1$ for all x. This function does not do much.

Let us see how much work we need to do in order to apply the Second Derivative Test. Using the quotient rule to differentiate and then simplifying, we get

$$f''(x) = \frac{-2x^3 + 18x^2 + 96x - 96}{(x^2 + 16)^3}.$$

Replacing x with -2, we have

$$f''(-2) = \frac{16 + 72 - 192 - 96}{(4 + 16)^3} = \frac{-200}{(20)^3} = -\frac{1}{40} < 0.$$

Since $f''(-2) < 0$, applying the Second Derivative Test we conclude that $f(x)$ has a local maximum value at $x = -2$.

Replacing x with 8, we get

$$f''(8) = \frac{-2(8)^3 + 18(8)^2 + 96(8) - 96}{(64 + 16)^3} = \frac{800}{(80)^3} = \frac{1}{640} > 0.$$

Since $f''(8) > 0$, applying the Second Derivative Test we conclude that $f(x)$ has a local minimum value at $x = 8$.

Exercises

1. Consider the function

$$f(x) = \frac{x^2 - x + 13}{x^2 + 9}.$$

Find the critical values of $f(x)$. Determine which of these critical values give a local maximum and which give a local minimum value for $f(x)$.

2. Suppose $f(x)$ is a continuous function which satisfies all the following conditions.

a) $f'(0) = DNE$, $f'(2) = 0$, $f'(4) = DNE$, $f'(6) = 0$, $f''(0) = DNE$, $f''(2) > 0$, $f''(4) = DNE$, and $f''(6) = 0$.

b) $f'(x) > 0$ for $x < 0$, for $2 < x < 4$, for $4 < x < 6$, and for $x > 6$. Also $f'(x) < 0$ for $0 < x < 2$.

c) $f''(x) > 0$ for $x < 0$, for $0 < x < 2$, for $2 < x < 4$ and for $x > 6$. Also $f''(x) < 0$ for $4 < x < 6$.

Find the critical values for $f(x)$. Determine if the function $f(x)$ has a local maximum value or a local minimum value at these critical values using the Second Derivative Test. For which values of x is the graph of $f(x)$ concave up?

3. Suppose $f(x)$ is a continuous function for all x with the following properties.

a) $f'(-2) = 0$, $f'(1) = DNE$, $f'(4) = DNE$, $f''(-2) < 0$, $f''(1) = DNE$, and $f''(4) = DNE$.

b) $f'(x) > 0$ for $x < -2$ and also for $1 < x < 4$. Also $f'(x) < 0$ for $-2 < x < 1$ and also for $x > 4$.

c) $f''(x) > 0$ for $1 < x < 4$ and for $x > 4$. Also $f''(x) < 0$ for $x < -2$ and for $-2 < x < 1$.

Sketch a possible graph of $f(x)$ for $-5 < x < 6$. Which values of x are critical points? Use the Second Derivative Test to test each of the critical points to determine if $f(x)$ has a local maximum value or a local minimum value at the critical point. For what values of x is the graph of $f(x)$ concave up?

4. Suppose $f(x)$ is a continuous function defined on the closed interval $-3 \le x \le 5$. Suppose $f(x)$ satisfies all the following conditions for $-3 \le x \le 5$.

a) $f'(-3) > 0$, $f'(-1) = 0$, $f'(2) = DNE$, $f'(4) = DNE$, $f'(5) < 0$, $f''(-1) < 0$, $f''(2) = DNE$, and $f''(4) = DNE$.

b) $f'(x) > 0$ for $-3 < x < -1$ and for $2 < x < 4$. Also $f'(x) < 0$ for $-1 < x < 2$ and for $4 < x < 5$.

c) $f''(x) > 0$ for $4 < x < 5$. Also $f''(x) < 0$ for $-3 < x < -1$, for $-1 < x < 2$, and for $2 < x < 4$.

Determine for what values of x the function $f(x)$ has a local maximum value and for what values of x the function $f(x)$ has a local minimum value. For what values of x does the graph of $f(x)$ have an inflection point? Sketch a possible graph for $f(x)$ in the interval $-3 \le x \le 5$.

5. Suppose $f(x)$ is a continuous function for all real numbers x with all the following properties.

a) $f'(-3) = DNE$, $f'(1) = 0$, $f'(4) = DNE$, $f''(-3) = DNE$, $f''(1) = 0$ and $f''(4) = DNE$.

b) $f'(x) > 0$ for $-3 < x < 1$ and for $1 < x < 4$. Also $f'(x) < 0$ for $x < -3$ and for $x > 4$.

c) $f''(x) > 0$ for $1 < x < 4$. Also $f''(x) < 0$ for $x < -3$ and for $-3 < x < 1$, and for $x > 4$.

For what values of x does the function $f(x)$ have a local maximum value and for what values does it have a local minimum? Do the problem once using the First Derivative Test and then do it again using the Second Derivative Test. Sketch a possible graph of $f(x)$.

6. Define the function $f(x)$ by

$$f(x) = \begin{cases} x^2 + 6x + 9 & x < 0 \\ -x^2 + 9 & 0 \le x < 3 \\ -x^2 + 12x - 27 & 3 \le x < 6 \\ x^2 - 12x + 45 & x \ge 6. \end{cases}$$

Find $f'(x)$ and $f''(x)$. Note that it is possible to see for example that $f'(0) = DNE$ and $f''(0) = DNE$. Find all the critical points for $f(x)$ For what values of x is $f(x)$ increasing and for what values of x is $f(x)$ decreasing? Use the first derivative test to determine which critical points give a local maximum value and which give a local minimum value. For what values of x does the graph of $f(x)$ have an inflection point?

6369 L'Hospital's Rule

At the beginning of our study of calculus we defined the derivative of a function using limits. We then computed some derivatives using limits. Let us review such a limit problem.

Example 1. Evaluate the limit

$$\lim_{x \to 6} \frac{x^2 - 8x + 12}{x^2 - 2x - 24}.$$

Solution. We have an easy method for evaluating limits. When the function is continuous we are able to evaluate the limit by substituting in the value. Let us try to evaluate this limit by substitution. Substituting $x = 6$ into this function gives

$$\frac{(6)^2 - 8(6) + 12}{(6)^2 - 2(6) - 24} = \frac{0}{0}.$$

First, $0/0$ is undefined. We did not get a number when we substituted. This function is not continuous at $x = 6$. The easy method (substitution) for finding the limit did not work. The expression $0/0$ is called "an indeterminate form". When substitution yields the indeterminate form $0/0$ we learned another method for finding the limit. Let us apply that method to finding this limit. Factoring the numerator and denominator, we rewrite the limit as

$$\lim_{x \to 6} \frac{(x - 6)(x - 2)}{(x - 6)(x + 4)}.$$

Now

$$\frac{(x - 6)(x - 2)}{(x - 6)(x + 4)} = \frac{x - 2}{x + 4} \text{ if } x \neq 6 \text{ and } x \neq -4.$$

Recall that in order to find the limit as x approaches 6 we use values of x that get closer and closer to 6, but never use the value $x = 6$. This means that

$$\frac{(x - 6)(x - 2)}{(x - 6)(x + 4)} = \frac{x - 2}{x + 4}$$

for all values of x which we must consider in finding the limit as x approaches 6. Therefore

$$\lim_{x \to 6} \frac{(x - 6)(x - 2)}{(x - 6)(x + 4)} = \lim_{x \to 6} \frac{x - 2}{x + 4}.$$

211

Let us try finding the limit $\lim\limits_{x \to 6} \dfrac{x - 2}{x + 4}$ using the easy method. Substituting 6 for x gives

$$\lim_{x \to 6} \frac{x - 2}{x + 4} = \frac{4}{10} = \frac{2}{5}.$$

It follows that

$$\lim_{x \to 6} \frac{x^2 - 8x + 12}{x^2 - 2x - 24} = \frac{2}{5}.$$

This solution depends upon being able to factor both the numerator and denominator of the fraction.

Example 2. Find the limit

$$\lim_{x \to 0} \frac{\sin x}{e^{2x} - 1}.$$

Solution. First, try the easy method to evaluate this limit. Substituting $x = 0$ gives $0/0$. This is the indeterminate form. When we get an indeterminate form the limit may exist or the limit may not exist. Following the solution of Example 1 we would next try to factor the numerator $\sin x$ and the denominator $e^{2x} - 1$. There is no way to factor these expressions. In order to find this limit we need a different method. Our new method is to apply the following theorem.

L'Hospital's Rule. Suppose $f(x)$ and $g(x)$ are differentiable and therefore continuous near $x = c$ and also at $x = c$. Suppose $g'(x) \neq 0$ for x near c and $x \neq c$. Suppose that

a) $\lim\limits_{x \to c} f(x) = 0$,   b) $\lim\limits_{x \to c} g(x) = 0$, and   c) $\lim\limits_{x \to c} \dfrac{f'(x)}{g'(x)} = L$,

then $\lim\limits_{x \to c} \dfrac{f(x)}{g(x)} = L$.

The proof of this theorem as stated here is relatively easy. The proof uses the Mean Value Theorem. Other cases of the theorem will be stated later.

Let us apply L'Hospital's Rule to find the limit in Example 2. We must show that a), b), and c) are all true.

$$\lim_{x \to 0} (\sin x) = 0 \qquad \lim_{x \to 0} e^{2x} - 1 = 0.$$

Since these functions are continuous we evaluated these two limits using the easy method. We simply substituted 0 for x. If $f(x) = \sin x$, then $f'(x) = \cos x$. If $g(x) = e^{2x} - 1$, then $g'(x) = 2e^{2x}$.

$$\lim_{x \to 0} \frac{f'(x)}{g'(x)} = \lim_{x \to 0} \frac{\cos x}{2e^{2x}} = \frac{1}{2}.$$

We find this limit by substituting zero for x. If we replace $f(x)$ with $\sin x$, $g(x)$ with $e^{2x} - 1$, and $c = 0$ in L'Hospital rule, we get the true statement. If $\lim_{x \to 0} \sin x = 0$, $\lim_{x \to 0} e^{2x} - 1 = 0$, and $\lim_{x \to 0} \frac{\cos x}{2e^{2x}} = \frac{1}{2}$, then $\lim_{x \to 0} \frac{\sin x}{e^{2x} - 1} = \frac{1}{2}$.

Since all three conditions in the hypothesis are satisfied, we conclude using L'Hospitals Rule that

$$\lim_{x \to 0} \frac{\sin x}{e^{2x} - 1} = \frac{1}{2}.$$

Example 3. Evaluate the limit $\lim_{x \to 1} \dfrac{\sin(x - 1)}{\ln x}$.

Solution. First, we hope this is a continuous function so that we can use the easy method (substitution) in order to find the limit. But substituting in $x = 1$ gives $0/0$, the indeterminate form. This means that the function $[\sin(x - 1)]/[\ln x]$ is not continuous at $x = 1$. Check to see if the conditions required in order to be able to apply L'Hospital's Rule are satisfied.

$$\lim_{x \to 1} [\sin(x - 1)] = 0 \quad \text{and} \quad \lim_{x \to 1} [\ln x] = 0.$$

If $f(x) = \sin(x-1)$, then $f'(x) = \cos(x-1)$. If $g(x) = \ln x$, then $g'(x) = 1/x$.

$$\frac{f'(x)}{g'(x)} = \frac{\cos(x - 1)}{1/x} = x \cos(x - 1).$$

We use the easy method (substitution) to find the following limit.

$$\lim_{x \to 1} \frac{f'(x)}{g'(x)} = \lim_{x \to 1} [x \cos(x - 1)] = 1.$$

The three conditions in order to be able to apply L'Hospital Rule are satisfied. It follows that

$$\lim_{x \to 1} \frac{\sin(x - 1)}{\ln x} = 1.$$

Example 4. Evaluate the limit

$$\lim_{x \to 0} \frac{e^{3x} - 1 - 3x}{x \sin(4x)}.$$

Solution. First, we hope that this is a continuous function at $x = 0$ and so we can find this limit by substituting. However, substituting $x = 0$ gives $0/0$, the well known indeterminate form. The function is not continuous. Substituting zero for x does not give the value of the limit. Let us check the conditions that must be satisfied in order to be able to apply L'Hospital's Rule.

$$\lim_{x \to 0} [e^{3x} - 1 - 3x] = 0 \text{ and } \lim_{x \to 0} [x \sin(4x)] = 0.$$

If $f(x) = e^{3x} - 1 - 3x$, then $f'(x) = 3e^{3x} - 3$. If $g(x) = x \sin(4x)$, then $g'(x) = \sin(4x) + 4x \cos(4x)$. When we try to find the limit

$$\lim_{x \to 0} \frac{3e^{3x} - 3}{\sin(4x) + 4x \cos(4x)}$$

by substituting $x = 0$, we get $0/0$, the indeterminate form. What to do? At first it seems like L'Hospital's rule has failed us.

Forget the original problem for the moment. Let us concentrate on finding this new limit. Let us try to use L'Hospital's Rule to find the limit

$$\lim_{x \to 0} \frac{3e^{3x} - 3}{\sin(4x) + 4x \cos(4x)}.$$

Check the conditions that must be satisfied in order to be able to apply L'Hospital's rule.

$$\lim_{x \to 0} [3e^{3x} - 3] = 0 \text{ and } \lim_{x \to 0} [\sin(4x) + 4x \cos(4x)] = 0.$$

If $f(x) = 3e^{3x} - 3$, then $f'(x) = 9e^{3x}$. If $g(x) = \sin(4x) + 4x \cos(4x)$, then $g'(x) = 8 \cos(4x) - 16x \sin(4x)$. Next, we find the limit:

$$\lim_{x \to 0} \frac{f'(x)}{g'(x)} = \lim_{x \to 0} \frac{9e^{3x}}{8 \cos(4x) - 16x \sin(4x)} = \frac{9}{8}.$$

214

We were able to evaluate this limit by substituting $x = 0$. The three conditions for L'Hospital's Rule are satisfied and we can conclude that

$$\lim_{x \to 0} \frac{3e^{3x} - 3}{\sin(4x) + 4x\cos(4x)} = \frac{9}{8}.$$

This limit is the limit we needed in order to be able to apply L'Hospital's rule to the original problem. Since we now know the value of this limit we can return to consideration of the original problem. The three conditions we need in order to apply L'Hospital's Rule to the original problem are true. We conclude that

$$\lim_{x \to 0} \frac{e^{3x} - 1 - 3x}{x\sin 4x} = \frac{9}{8}.$$

Example 5. Evaluate the limit $\lim_{x \to +\infty} \dfrac{\ln x}{x^2}$.

Solution. First, since $\lim_{x \to +\infty} \ln x = +\infty$ and $\lim_{x \to +\infty} x^2 = +\infty$, we get ∞/∞. This is not our usual indeterminate form which is $0/0$. However, this is an indeterminate form for which L'Hospital's Rule is also true.

L'Hospital's Rule Case II. If $\lim_{x \to c} f(x) = +\infty$, $\lim_{x \to c} g(x) = +\infty$, and $\lim_{x \to c} \dfrac{f'(x)}{g'(x)} = L$, then $\lim_{x \to c} \dfrac{f(x)}{g(x)} = L$.

This case II is difficult to prove. This case is true when c is a finite number and is also true when c is replaced by $+\infty$. The proof can only be done in a more advanced mathematics course. We can apply L'Hospital's Rule case II to the problem in Example 5. We have already seen that

$$\lim_{x \to +\infty} \ln(x) = +\infty \text{ and } \lim_{x \to +\infty} x^2 = +\infty.$$

If $f(x) = \ln(x)$, then $f'(x) = 1/x$. If $g(x) = x^2$, then $g'(x) = 2x$.

$$\lim_{x \to +\infty} \frac{f'(x)}{g'(x)} = \lim_{x \to +\infty} \frac{1/x}{2x} = \lim_{x \to +\infty} \frac{1}{2x^2} = 0.$$

Applying L'Hospital's Rule Case II, we conclude that

$$\lim_{x \to +\infty} \frac{\ln x}{x^2} = 0.$$

Example 6. The solution of a certain logistic initial value problem can be written as $y = \dfrac{10 + 24e^{5x}}{25 + 6e^{5x}}$. Evaluate the limit $\lim_{x \to +\infty} \dfrac{10 + 24e^{5x}}{25 + 6e^{5x}}$.

Solution. Recall that $\lim_{x \to +\infty} e^{5x} = +\infty$. It follows that

$$\lim_{x \to +\infty} (10 + 24e^{5x}) = +\infty \text{ and } \lim_{x \to +\infty} (25 + 6e^{5x}) = +\infty.$$

Therefore, $\lim_{x \to +\infty} \dfrac{10 + 24e^{5x}}{25 + 6e^{5x}} = \dfrac{+\infty}{+\infty}$. This is an indeterminate form as expressed in L'Hospital's Rule Case II. If $f(x) = 10 + 24e^{5x}$, then $f'(x) = 120e^{5x}$. If $g(x) = 25 + 6e^{5x}$, then $g'(x) = 30e^{5x}$.

$$\lim_{x \to +\infty} \frac{f'(x)}{g'(x)} = \lim_{x \to +\infty} \frac{120e^{5x}}{30e^{5x}} = \lim_{x \to +\infty} (4) = 4.$$

It follows applying L'Hospital's Rule Case II that

$$\lim_{x \to +\infty} \frac{10 + 24e^{5x}}{25 + 6e^{5x}} = 4.$$

In stating L'hospital's Rule we have assumed that the c in $\lim_{x \to c}$ is a finite number and not $+\infty$. However, L'Hospital's Rule Case I and Case II are both true if we replace c with $+\infty$ or if we replace c with $-\infty$. These cases can also be handled by making a substitution of variable as indicated in the following example.

Example 7. Evaluate the limit

$$\lim_{x \to +\infty} \frac{x^{-2/3} + 5x^{-4/3}}{3x^{-1} + 4x^{-1/3}}.$$

Solution. First note that

$$\lim_{x \to +\infty} (x^{-2/3} + 5x^{-4/3}) = 0 \text{ and } \lim_{x \to +\infty} (3x^{-1} + 4x^{-1/3}) = 0.$$

Instead of applying L'Hospital's Rule directly let us make a change of variable. Let $y = 1/x$, then

$$\lim_{x \to +\infty} (x^{-2/3} + 5x^{-4/3}) = \lim_{y \to 0+} (y^{2/3} + 5x^{4/3})$$

$$\lim_{x \to +\infty} (3x^{-1} + 4x^{-1/3}) = \lim_{y \to 0+} (3y + 4y^{1/3}).$$

The problem is restated as

$$\lim_{y \to 0+} \frac{y^{2/3} + 5y^{4/3}}{3y + 4y^{1/3}} = \lim_{y \to 0+} \frac{y^{1/3} + 5y}{3y^{2/3} + 4}.$$

Substituting $y = 0$ we get $0/4 = 0$. This is no longer an indeterminant form

$$\lim_{x \to +\infty} \frac{x^{-2/3} + 5x^{-4/3}}{3x^{-1} + 4x^{-1/3}} = 0.$$

Example 8. Consider $\lim_{x \to 0} \dfrac{4e^{5x}}{2 + \sin 4x}$.

Solution. We easily evaluate this limit since the function is continuous. It is

$$\lim_{x \to 0} \frac{4e^{5x}}{2 + \sin 4x} = \frac{4}{2} = 2.$$

Suppose we make a mistake and apply L'Hospital's Rule. If $f(x) = 4e^{5x}$, then $f'(x) = 20e^{5x}$. If $g(x) = 2 + \sin 4x$, then $g'(x) = 4\cos 4x$.

$$\lim_{x \to 0} \frac{20e^{5x}}{4\cos 4x} = 5.$$

This mistake might cause us to think that the value of $\lim_{x \to 0} \dfrac{4e^{5x}}{2 + \sin 4x}$ is 5.

L'Hospital's Rule is sometimes used to find exotic limits. This is usually done in a way that takes advantage of the continuity of the function e^x. The most popular such function is x^x. In case you meet a person in a dark alley late at night and he points a gun at you and says "I want to discuss the function x^x" you need to be prepared. First, $f(x) = e^x = e^{x \ln x}$. This defines x^x for $x > 0$. No one tries to define x^x for $x < 0$. Note that $f(0.1) = 0.794$, $f(0.01) = 0.955$, $f(0.001) = 0.933$, and $f(0.0001) = 0.999$. This makes it clear that $\lim\limits_{x \to 0+} x^x = 1$. The usual proof of this is as follows. First

$$\lim_{x \to 0+} (x \ln x) = \lim_{x \to 0+} \frac{\ln x}{(1/x)} \overset{H}{=} \lim_{x \to 0+} \frac{1/x}{-(1/x^2)} = \lim_{x \to 0+} (-x) = 0.$$

Then we use the fact that the exponential function is continuous to say

$$\lim_{x \to 0+} x^x = \lim_{x \to 0+} e^{x \ln x} = \exp(\lim_{x \to 0+} x \ln x) = e^0 = 1.$$

We can define a function $g(x)$ as follows

$$g(x) = \begin{cases} x^x & \text{for } x > 0 \\ 1 & x = 0. \end{cases}$$

The function $g(x)$ is continuous for $x \geq 0$. Also

$$g'(x) = e^{x(\ln x)}(1 + \ln x) = x^x(1 + \ln x) \text{ for } x > 0.$$

Note that $\lim\limits_{x \to 0+} g'(x) = -\infty$.

Exercises

Evaluate each of the following limits.

1. $\lim\limits_{x \to 0} \dfrac{e^{5x} - 1}{\sin 3x}$

2. $\lim\limits_{x \to 0} \dfrac{\arcsin(x)}{x}$

3. $\lim\limits_{x \to 0} \dfrac{e^{5x} - 1 - 5x}{1 - \cos 4x}$

4. $\lim\limits_{x \to 3} \dfrac{x^3 - x^2 - 21x + 45}{x^3 - 8x^2 + 21x - 18}$

5. $\lim\limits_{x \to 2} \dfrac{\ln(x - 1)}{x - 2}$

6. $\lim\limits_{x \to 0} \dfrac{x \tan x}{e^{4x} - 1 - 4x}$

7. $\lim\limits_{x \to +\infty} \dfrac{\ln x}{\sqrt{x}}$

8. $\lim\limits_{x \to +\infty} \dfrac{3x^2 + 5x}{4x^2 + 7}$

9. $\lim\limits_{x \to 0+} \dfrac{e^{5x} - 5x}{(\sin 4x)}$

10. $\lim\limits_{x \to +\infty} \dfrac{x^{-2} + 12x^{-4/3}}{8x^{-5/3} + 2x^{-4/3}}$

11. The limit $\lim\limits_{x \to 0} \dfrac{4x + 10\cos 3x}{20e^{3x} + 5} = \dfrac{2}{5}$ is an easy limit. Make a mistake and assume L'Hospital's Rule applies, then what do you think the limit is?

6371 The Mean Value Theorem for Derivatives

Rolle's Theorem. Suppose $f(x)$ is a function such that

    1. $f(x)$ is continuous for $a \le x \le b$
    2. $f'(x)$ exists for $a < x < b$
    3. $f(a) = f(b) = 0$

Then there exists a number c with $a < c < b$ such that $f'(c) = 0$.

Proof. Case 1. Suppose $f(x) \equiv 0$, then $f'(x) \equiv 0$. In this case the number c can be any number in the interval $a < x < b$.

Case 2. Suppose there is some value of x such that $f(x) > 0$, then there must be a value of x say $x = c$ such that $f(x)$ has a local maximum value at $x = c$ with $a < c < b$. Since $f(x)$ has a local maximum at $x = c$ with $a < c < b$ and $f'(c)$ exists, then $f'(c) = 0$.

Case 3. Suppose there is some value of x such that $f(x) < 0$, then there must be a value of x say $x = c$ such that $f(x)$ has a local minimum value at $x = c$ with $a < c < b$. Since $f(x)$ has a local minimum at $x = c$ with $a < c < b$ and $f'(c)$ exists, then $f'(c) = 0$.

It follows that there is always one or more values of c such that $f'(c) = 0$ and $a < c < b$.

Rolle's theorem is a special case of the Mean Value Theorem for Derivatives. The main use we make of Rolle's theorem is to use it in the proof of the following mean value theorem.

The Mean Value Theorem for Derivatives. If $f(x)$ is a function such that

    1. $f(x)$ is continuous for $a \le x \le b$.
    2. $f'(x)$ exists for $a < x < b$,

then there is a number c with $a < c < b$ such that

$$f'(c) = \frac{f(b) - f(a)}{b - a}.$$

Proof. First, we define the function $h(x)$ as

$$h(x) = f(x) - f(a) - \frac{f(b) - f(a)}{b - a}(x - a).$$

Since $f(x)$ and $(x - a)$ are continuous for $a \leq x \leq b$ it follows that $h(x)$ is continuous for $a \leq x \leq b$. The derivative of $h(x)$ is

$$h'(x) = f'(x) - \frac{f(b) - f(a)}{b - a}.$$

Hence, the derivative $h'(x)$ exists for $a < x < b$. Evaluating $h(x)$ when $x = a$ and $x = b$, we get

$$h(a) = f(a) - f(a) - \frac{f(b) - f(a)}{b - a}(a - a) = 0$$

$$h(b) = f(b) - f(a) - \frac{f(b) - f(a)}{b - a}(b - a) = 0$$

We have shown that $h(x)$ satisfies all three parts of the hypothesis of Rolle's Theorem. It follows that $h(x)$ satisfies the conclusion of Rolle's Theorem that is, there exists a number c with $a < c < b$ such that $h'(c) = 0$. This says

$$f'(c) - \frac{f(b) - f(a)}{b - a} = 0$$

$$f'(c) = \frac{f(b) - f(a)}{b - a}.$$

Geometrically the Mean Value Theorem for Derivatives says that there is a number c such that the slope of the tangent line to the curve $y = f(x)$ at $x = c$ is the same as the slope of the secant line connecting the points $(a, f(a))$ and $(b, f(b))$.

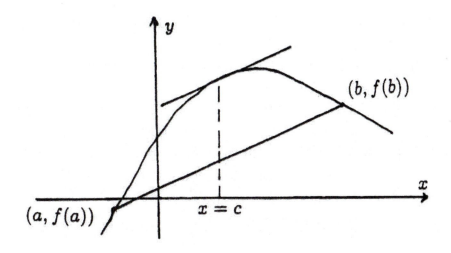

221

To say that the Mean Value Theorem for Derivatives is true means we can replace $f(x)$ in the Mean Value Theorem for Derivatives with a specific function and get a true statement. These statements are usually called implications. There are two parts to an implication. The phrase that comes after if and before then is called the hypothesis and the phrase that comes after then is called the conclusion of the implication. Both the hypothesis and conclusion must be complete thoughts and so have a subject and verb. In our reasoning in mathematics we are usually interested in implications that are true and also have a true conclusion. This means that the hypothesis must also be true. Our usual situation is that we have a true implication. We check to see that the hypothesis of the implication is true. We can then conclude that the conclusion is true. Then we replace $f(x)$ in the Mean Value Theorem for Derivatives with a specific function we get a true statement. However, we only get a useful true statement if the hypothesis is true. If the hypothesis of the Mean Value Theorem for Derivatives is true, then the conclusion is true. The conclusion of the Mean Value Theorem for Derivatives is the formula $f(b) - f(a) = f'(c)(b - a)$. If we consider a function $f(x)$ for which both parts of the hypothesis of the Mean Value Theorem for Derivatives are true, then the formula $f(b) - f(a) = f'(c)(b - a)$ is true since it is the conclusion of the Mean Value Theorem for Derivatives.

In order to gain understanding of the Mean Value Theorem for Derivatives we work exercises like the following. In these exercises we are simply trying to see exactly what are the hypothesis of the Mean Value Theorem for Derivatives and what are the conclusions of the Mean Value Theorem for Derivatives. Also to see that all are true. We are not doing anything that is really useful.

Example 1. Let $f(x) = x^3 + 2x^2 + 5x$ for $0 \leq x \leq 3$. Verify that $f(x)$ satisfies the hypotheses of the Mean Value Theorem for Derivatives for $0 \leq x \leq 3$. Second, find all values of c that satisfy the conclusion of the Mean Value Theorem for Derivatives.

Solution. Verify means to actually check to see that the statements are true. The first hypotheses is: the function $f(x) = x^3 + 2x^2 + 5x$ is continuous for $0 \leq x \leq 3$. This is clearly true. The second hypotheses is that the function $f'(x) = 3x^2 + 4x + 5$ exists and is the derivative of $f(x)$ for $0 \leq x \leq 3$. This is clearly true. We have verified that the hypothesis of the Mean Value

Theorem for Derivatives are satisfied. The values of c which satisfy the conclusion of the Mean Value Theorem for Derivatives are the numbers c that are solutions of the following equation.

$$f'(c) = \frac{f(3) - f(0)}{3 - 0}$$

$$3c^2 + 4c + 5 = \frac{(3)^3 + 2(3)^2 + 5(3) - (0)}{3 - 0} = 20$$

$$3c^2 + 4c - 15 = 0$$

$$(3c - 5)(c + 3) = 0$$

$$c = 5/3 \text{ and } c = -3.$$

At first we may think there are two values of c. However, c must satisfy $0 < c < 3$. The value $c = -3$ does not satisfy this condition. It follows that the only value of c is $c = 5/3$ since $0 < 5/3 < 2$.

Suppose $y = g(t)$ denotes the position of a particle on the y axis at time t, then $g'(t)$ is the instantaneous velocity at time t. For the position function $g(t)$ the Mean Value Theorem for Derivatives says

$$\frac{g(b) - g(a)}{b - a} = g'(c) \text{ for } a < c < b.$$

The expression $\frac{g(b) - g(a)}{b - a}$ is the average velocity of the particle over the time interval $a \le t \le b$. The Mean Value Theorem for Derivatives says that there is some instant in time when the instantaneous velocity $g'(t)$ is equal to this average velocity.

Example 2. Let $f(x) = x^{2/3} + 1$ for $-1 \le x \le 8$. Show that there is no number c which satisfies the conclusion of the Mean Value Theorem for Derivatives for $-1 \le x \le 8$. Why does this not contradict the Mean Value Theorem for Derivatives?

Solution. First, $f(-1) = (-1)^{2/3} + 1 = 2$ and $f(8) = (8)^{2/3} + 1 = 5$. Since $f'(x) = (2/3)x^{-1/3}$, the value of c is the solution of the equation

$$\frac{2}{3}c^{-1/3} = \frac{5 - 2}{8 - (-1)} = \frac{1}{3}.$$

$$c^{-1/3} = \frac{1}{2}$$

$$c = 8.$$

We cannot say $c = 8$ because it is not true that $-1 < 8 < 8$. There is not a value of c which satisfies the conclusion of the Mean Value Theorem for Derivatives. The conclusion of the Mean Value Theorem for Derivatives for $f(x) = x^{2/3} + 1$ for $-1 \leq x \leq 8$ is false.

The two hypothesis of the Mean Value Theorem for Derivatives for this function are:

1. The function $f(x) = x^{2/3} + 1$ is continuous for $-1 \leq x \leq 8$.

2. $f'(x) = (2/3)x^{-1/3}$ exists for $-1 < x < 8$.

The first hypothesis is true. The second hypothesis is false since $f'(0) = DNE$. Even though the conclusion of the Mean Value Theorem for Derivatives for the example is false, it does not contradict the Mean Value Theorem for Derivatives because the hypothesis is also false.

Example 3. Let $f(x) = 2|x - 2| + |x - 5|$ for $0 \leq x \leq 6$. State the two hypothesis and the conclusion of the Mean Value Theorem for Derivatives for this function.

Solution. We can rewrite the function $f(x)$ as

$$f(x) = \begin{cases} -3x + 9 & 0 \leq x < 2 \\ x + 1 & 2 \leq x < 5 \\ 3x - 9 & 5 \leq x \leq 6. \end{cases}$$

A sketch of the graph of $f(x)$ is

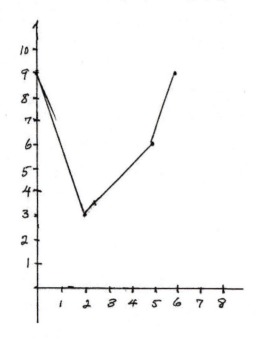

The hypothesis are

1. $f(x) = 2|x - 2| + |x - 5|$ is continuous for $0 \leq x \leq 6$.

2. The derivative for $0 < x < 6$ is given by

$$f'(x) = \begin{cases} -3 & 0 < x < 2 \\ DNE & x = 2 \\ 1 & 2 < x < 5 \\ DNE & x = 5 \\ 3 & 5 < x < 6. \end{cases}$$

The conclusion is

$$f'(c) = \frac{f(6) - f(0)}{6 - 0} = \frac{9 - 9}{6 - 0} = 0 \text{ for } 0 < c < 6.$$

Clearly there is no value of c such that $f'(c) = 0$.

The first hypothesis is true. The second hypothesis is false. The conclusion is false.

The primary application of the Mean Value Theorem for Derivatives is to prove theorems.

Example 4. Show for all a and b that

$$|\sin(b) - \sin(a)| \leq |b - a|.$$

Solution. Let $f(x) = \sin x$ then $f(x)$ is continuous for any interval $a \leq x \leq b$. Also $f'(x) = \cos x$ exists for any interval $a < x < b$. Both parts of the hypothesis of the Mean Value Theorem for Derivatives are true and therefore the conclusion is true. The conclusion of the Mean Value Theorem for Derivatives says that for any interval $a \leq x \leq b$ there is a number c such that

$$\cos(c) = \frac{\sin(b) - \sin(a)}{b - a},$$

with $a < c < b$. Taking absolute values gives

$$\left| \frac{\sin(b) - \sin(a)}{b - a} \right| = |\cos(c)|.$$

We know $|\cos(c)| \le 1$. Thus we have

$$\left| \frac{\sin(b) - \sin(a)}{b - a} \right| \le 1.$$

It follows that

$$|\sin(b) - \sin(a)| \le |b - a|.$$

Theorem 3. Suppose $f(x)$ is a function such that

    1) $f(x)$ is continuous for $a \le x \le b$
    2) $f'(x)$ exists for $a < x < b$
    3) $\lim\limits_{x \to a+} f'(x)$ exists.

Then $\lim\limits_{x \to a+} \dfrac{f(x) - f(a)}{x - a} = \lim\limits_{x \to a+} f'(x)$.

Proof. Statements 1 and 2 say that the hypothesis of the Mean Value Theorem for Derivatives is true. Therefore, the conclusion is true, that is, there exists a number θ such that

$$\frac{f(x) - f(a)}{x - a} = f'(\theta) \text{ where } a < \theta < x.$$

To say $x \to a+$ requires that $\theta \to a+$ since $a < \theta < x$. It is given that $\lim\limits_{x \to a+} f(x)$ exists. This means that

$$\lim_{x \to a+} f'(\theta) = \lim_{x \to a+} f'(x)$$

exists. It follows that

$$\lim_{x \to a+} \frac{f(x) - f(a)}{x - a} = \lim_{x \to a+} f'(\theta) = \lim_{x \to a+} f'(x).$$

We used this fact when discussing one-sided derivatives.

Exercises

1. Let $f(x) = x^3 - 7x^2 + 15x$ for $1 \le x \le 5$. State both parts of the hypothesis and also the conclusion of the Mean Value Theorem for Derivatives for

this function and this interval. Are each of the two parts of the hypothesis true? Is the conclusion true? Second, find all values of c that satisfy the conclusion of the Mean Value Theorem for Derivatives. Ans. The value of c is $c = 11/3$.

2. Let $f(x) = \sqrt{x+1} - x$ for $0 \le x \le 3$. State both parts of the hypothesis and also the conclusion of the Mean Value Theorem for Derivatives for this function. Are each of the two parts of the hypothesis true? Is the conclusion true? Second, find all values of c that satisfy the conclusion of the Mean Value Theorem for Derivatives.

3. Let $f(x) = \dfrac{x+4}{x-4}$. Show that there is no value of c with $0 < c < 5$ such that

$$f'(c) = \frac{f(5) - f(0)}{5 - 0}.$$

Why does this not contradict the Mean Value Theorem for Derivatives?

4. Let $f(x) = \dfrac{x-1}{x+3}$ for $1 \le x \le 6$. Verify that $f(x)$ satisfies the hypothesis for the Mean Value Theorem for Derivatives for $1 \le x \le 6$. Second, find all values of c that satisfy the conclusion of the Mean Value Theorem for Derivatives.

5. Consider $f(x) = |x - 3| + x$ for $0 \le x \le 5$. Which of the hypothesis of the Mean Value Theorem for Derivatives is false for this function and this interval?

6. Let $f(x) = \begin{cases} -x^2 + 10x & 0 \le x < 2 \\ 16 & x = 2 \\ 2x^2 - 2x + 12 & 2 < x \le 8 . \end{cases}$

State both parts of the hypothesis and also the conclusion of the Mean Value Theorem for Derivatives for this function and the interval $0 \le x \le 8$. Are each of the parts of the hypothesis true? Is the conclusion true?

7. Let $f(x) = \begin{cases} x^2 + 1 & x < 0 \\ 1 & 0 \le x \le 2 \\ x^2 - 3 & x > 2 . \end{cases}$

State both parts of the hypothesis and also the conclusion of the Mean Value Theorem for Derivatives for this function and the interval $-1 \le x \le 5$. Are each of the parts of the hypothesis true? Is the conclusion true?

8. Let $f(x) = (25 - x^2)^{1/2}$ for $-3 \leq x \leq 5$. State both parts of the hypothesis of the Mean Value Theorem for Derivatives for this function and this interval. State the conclusion of the Mean Value Theorem for Derivatives for this function and this interval. Are each of the parts of the hypothesis true? Is the conclusion true?

9. Suppose $f(x)$ is a function such that

1. $f(x)$ is continuous for $a \leq x \leq b$.

2. $f'(x)$ exists for $a < x < b$.

3. $\lim\limits_{x \to a+} f'(x) = +\infty$.

then $\lim\limits_{x \to a+} \dfrac{f(x) - f(a)}{x - a} = +\infty$.

Prove that this statement is true using the same method that was used to prove Theorem 3. Note that the conclusion can also be stated as $f'(a) = +\infty$.

6373 Linearization

Given the function $f(x)$ recall the definition of the derivative of $f(x)$ at the point $x = c$.

$$f'(c) = \lim_{x \to c} \frac{f(x) - f(c)}{x - c}.$$

Since $\lim_{x \to c} \dfrac{f(x) - f(c)}{x - c} = f'(c)$, it follows that for values of x near c, we have

$$\frac{f(x) - f(c)}{x - c} \simeq f'(c).$$

Stated another way, if x is any number such that $|x - c| < d$ for d sufficiently small, then

$$\frac{f(x) - f(c)}{x - c} \simeq f'(c).$$

Thus, for x such that $|x - c| < d$, we have

$$f(x) - f(c) \simeq f'(c)(x - c)$$

or

$$f(x) \simeq f(c) + f'(c)(x - c).$$

Definition. Linearization. Given a function $f(x)$ the linear approximation, $L(x)$, of $f(x)$ at $x = c$ is

$$L(x) = f(c) + f'(c)(x - c).$$

The function $L(x)$ is also called the linearization of $f(x)$ at $x = c$. We have just shown that if $|x - c|$ is small, then $L(x)$ is approximately equal to $f(x)$. This is written as $f(x) \simeq L(x)$. We sometimes use the notation $L_c(x)$ instead of the simpler $L(x)$ in order to include c as part of the notation. Linearizations are often used in physics and engineering. In analyzing the properties of a function there is often a need to simplify the function by replacing it with its linear approximation.

We are now going to consider the problem of finding the linearization of a given function. Once we have found the linearization $L_c(x)$ of $f(x)$ at

$x = c$ we will use $L_c(x)$ to approximate $f(x)$ for a value of x near c. We do this in order to remind us that $L_c(x) \simeq f(x)$ for $|x - c|$ small. This is an important fact about linearization that we need to remember.

Example 1. Let $f(x) = \sqrt{3x + 7}$. Find the linear approximation of $f(x)$ at $x = 6$ and use it to find an approximate value of $\sqrt{24.4}$.

Solution. First note that $f(6) = \sqrt{18 + 7} = 5$. Next $f'(x) = (3/2)(3x + 7)^{-1/2}$ and so $f'(6) = (3/2)(25)^{-1/2} = 3/10$. The linear approximation of $f(x)$ at $x = 6$ is
$$L(x) = 5 + (3/10)(x - 6).$$

What number must be substituted for x in order to make $3x + 7 = 24.4$? Solving this equation, we get $x = 5.8$. This says that $f(5.8) = \sqrt{24.4}$. Since 5.8 is nearly equal to 6 and $\sqrt{3x + 7} \simeq 5 + (3/10)(x - 6)$ for x near 6, we use $L(5.8)$ to approximate $f(5.8) = \sqrt{24.4}$.

$$L_6(5.8) = 5 + (3/10)(5.8 - 6) = 5 - (6/100) = 4.94.$$

Therefore, $\sqrt{24.4} \simeq 4.94$.

Example 2. Use linearization to find an approximate value of $(36.4)^{5/2}$.

Solution. Since we have the 5/2 power, we decide to consider the function $f(x) = x^{5/2}$, $x > 0$. We need a value of c that satisfies two conditions. The first condition is that the value of c needs to be near 36.4. The second condition is that the value of c needs to have an easily computed square root. Note that $c = 36$ satisfies both of these conditions. Let us find the linearization of $f(x) = x^{5/2}$ at $c = 36$. First, we need $f(36) = (36)^{5/2} = 7776$. Second, $f'(x) = (5/2)x^{3/2}$ and so $f'(36) = (5/2)(36)^{3/2} = 540$. The linearization of $f(x) = x^{5/2}$ at $c = 36$ is

$$L(x) = 7776 + 540(x - 36).$$

Next, we use the fact that $f(36.4) = (36.4)^{5/2}$ is approximately equal to $L(36.4)$.
$$L(36.4) = 7776 + 540(36.4 - 36) = 7992.$$

Using the linearization we say that $(36.4)^{5/2}$ is approximately equal to 7992.

Example 3. The function $y = f(x)$ is defined as follows. For a given value of x the corresponding value of y is a solution of the equation

$$3x^2 + y^2 - 6x + 4y = 21.$$

Also the function $y = f(x)$ is the solution such that when $x = 3$, then $y = 2$. Find the linearization of $f(x)$ at $c = 3$.

Solution. In order to find $\frac{dy}{dx} = f'(x)$ we need to use implicit differentiation. Taking the derivative of both sides of the given equation, we get

$$\frac{d}{dx}[3x^2 + y^2 - 6x + 4y] = \frac{d}{dx}[21]$$

$$6x + 2y\frac{dy}{dx} - 6 + 4\frac{dy}{dx} = 0$$

$$\frac{dy}{dx} = \frac{-6x + 6}{2y + 4}.$$

In order to find $f'(3)$ we substitute $x = 3$ and $y = 2$, into $\frac{dy}{dx}$.

$$f'(3) = \frac{-6(3) + 6}{2(2) + 4} = -\frac{3}{2}.$$

The linearization at $x = 3$ is

$$L(x) = 2 - (3/2)(x - 3).$$

We can use this linearization to find an approximate value of y corresponding to $x = 3.2$.

$$L(3.2) = 2 - (3/2)(3.2 - 3) = 1.7.$$

Example 4. Let $f(x) = \frac{\sin x + \cos x}{4 + \sin x}$. Find the linearization of $f(x)$ at $x = 0$. Use this linearization to find an approximate value of $f(0.04)$.

Solution. First, $f(0) = 1/4$. Next, the derivative of $f(x)$ is

$$f'(x) = \frac{4\cos x - 4\sin x - 1}{(4 + \sin x)^2}.$$

$$f'(0) = 3/16.$$

Using these values the linearization is

$$L(x) = \frac{1}{4} + \frac{3}{16}x.$$

The approximation is

$$L(0.04) = 0.25 + \frac{3}{16}(0.04) = 0.2575.$$

Exercises

1. Find the linearization of the function $f(x) = \sqrt{6x+4}$ at $c = 10$ and use it to approximate the numbers $\sqrt{62.8}$ and $\sqrt{64.9}$.

2. Find the linearization of the function $f(x) = x^{-3/2}$ at $c = 400$ and use it to approximate $(403)^{-3/2}$.

3. Use linear approximation to find an approximate value of $(4.06)^{3/2} + 5(4.06)^{-1/2}$.

4. Let $f(x) = \frac{x}{1+\sqrt{x}}$. Find the linear approximation of $f(x)$ at $x = 4$. Use this linear approximation to find an approximate value of $f(x)$ at $x = 4.3$.

5. The function $y = f(x)$ is defined as follows. For a given value of x the corresponding value of y is a solution of the equation

$$x^2 + y^3 - 6x + 2y = 3.$$

The function $y = f(x)$ is the solution whose graph passes through the point $(3, 2)$. Find the linearization of $f(x)$ at the point $c = 3$.

6. Find the linearization of the function $f(x) = \frac{\sin x}{2+\cos x}$ at $x = \pi/2$.

6375 Newton's Method

Suppose we want to find the solution of an equation. If we are given a quadratic equation, we can find an exact solution using the quadratic formula. For more complicated equations such as $2\sin x = x$ or $x^3 - 2x^2 - 7x + 8 = 0$ we do not have an easy method which gives an exact solution. For such equations we try to find an approximate solution. We would like to use a method which gives us the approximate solution correct to as many decimal places as we want. We are going to study a method of solution known as Newton's method. We choose to study this method now because it involves using the derivative.

Suppose we want to solve the equation $f(x) = 0$, but have no way to find the exact solution $x = r$ such that $f(r) = 0$. Suppose also that we have already found a number c which is relatively close to r, that is, $|r - c|$ is a somewhat small number. Starting with the number c, we find a better approximation to the exact solution r as follows. Below we will discuss methods for locating a number c which is somewhat close to the actual root r. The number c is called the initial guess. First, we find the linear approximation $L_c(x)$ of $f(x)$ at the point where $x = c$. Recall that for values of x near c, we have $f(x) \simeq L_c(x)$. It is easy to solve the equation $L_c(x) = 0$ since this is just a linear equation. If $f(x)$ is approximately equal to $L_c(x)$ for values of x near c, then the solution of the equation $L_c(x) = 0$ is approximately equal to the solution of the given equation $f(x) = 0$. Denote the solution of $L_c(x) = 0$ by $x = b$, then $L_c(b) = 0$. This says that the solution of $L_c(x) = 0$ is approximately equal to the solution of $f(x) = 0$. We expect that the new value b is closer to the actual solution of $f(x) = 0$ than is our initial value c, that is, $|f(b)| < |f(c)|$.

Recall the linear approximation of $f(x)$ at $x = c$ is

$(1+\sqrt{x})(1+\sqrt{x})$
$1+x+2\sqrt{x}$

$$L_c(x) = f(c) + f'(c)(x - c).$$

If $x = b$ is the value of x such that $L_c(b) = 0$, then

$$f(c) + f'(c)(b - c) = 0.$$

Solving for b, we get

$$b = c - \frac{f(c)}{f'(c)}.$$

233

If c is near a solution of the equation $f(x) = 0$, then b will be still nearer the solution. When we use this formula to improve our approximation to the solution of an equation we say that we are using Newton's method.

Example 1. Show that the equation $x^3 - 2x^2 - 7x + 8 = 0$ has a solution between $x = 3$ and $x = 4$. Find an approximate value of this solution using Newton's method.

Solution. Suppose we are going to graph the function $f(x) = x^3 - 2x^2 - 7x + 8$ by hand. For $x = 3$, we have $f(3) = (3)^3 - 2(3)^2 - 7(3) + 8 = -4$. When $x = 4$, we have $f(4) = 12$. We have found two points on the graph of $f(x) = x^3 - 2x^2 - 7x + 8$, namely $(3, -4)$ and $(4, 12)$. If we plot these two points and connect them with a smooth curve, we have part of the graph of the function $f(x) = x^3 - 2x^2 - 7x + 8$.

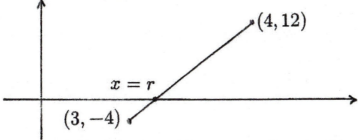

The graph of $f(x)$ crosses the x axis for a value of x, say $x = r$, such that $3 < r < 4$. This point on the graph has y coordinate zero. This point is $(r, 0)$. This means $f(r) = 0$. There is a solution of the equation $f(x) = 0$ for a value of x between $x = 3$ and $x = 4$.

As our first guess in order to apply Newton's method we can pick from the two values of x that we have already considered, namely $x = 3$ and $x = 4$. We should pick the one nearest the solution. Clearly 3 is closer to the actual solution than is 4 since -4 is nearer zero than 12. We find the linear approximation for $f(x)$ at $c = 3$.

$$f'(x) = 3x^2 - 4x - 7$$
$$f'(3) = 3(3)^2 - 4(3) - 7 = 8.$$

The linear approximation for $f(\tilde{x})$ at $x = 3$ is

$$L_3(x) = -4 + 8(x - 3).$$

The solution of the equation $-4 + 8(x - 3) = 0$ is $x = 3.5$. Since $f(3) = -4$ and $f(3.5) = 1.875$, it is clear that $x = 3.5$ is nearer to the exact solution of $x^3 - 2x^2 - 7x + 8 = 0$ than is $x = 3$. This is because $|1.875| < |-4|$.

234

Now that we have the approximation $x = 3.5$ we can try to get an even better approximation for the solution. We use the linear approximation to $f(x)$ again, but this time we find the approximation at $c = 3.5$.

$$f'(3.5) = 3(3.5)^2 - 4(3.5) - 7 = 15.75.$$

The linear approximation is

$$L_{3.5}(x) = 1.875 + 15.75(x - 3.5).$$

We next find the solution of the equation $L_{3.5}(x) = 0$. We solve

$$1.875 + 15.75(x - 3.5) = 0.$$

The solution is

$$x = 3.5 - \frac{1.875}{15.75} = 3.3810.$$

Since $f(3.3810) = 0.1194$ and $f(3.5) = 1.875$ it is clear that $x = 3.3810$ is nearer the exact solution of $x^3 - 2x^2 - 7x + 8 = 0$ than is $x = 3.5$. This is because $0.1194 < 1.875$.

Our best approximation for the solution so far is $x = 3.3810$. We could get one by finding the linear approximation at $c = 3.3810$ and using this to get a better value still for the solution of $x^3 - 2x^2 - 7x + 8 = 0$. When we perform the same series of calculations over and over it is called an iteration. Let us set up the procedure for approximating this solution as an iteration. Let x_1 denote our first guess for the solution, x_2 denote our second guess, x_3 our third guess, and continue in this manner. Note that x_2 is our first guess using the formula. Our first guess x_1 must be discovered somehow in order to start the calculations. We have now decided to call our first guess x_1 instead of c and our second guess x_2 instead of b. But looking back at our previous work we see that the formula for x_2 is

$$x_2 = x_1 - \frac{f(x_1)}{f'(x_1)}.$$

Once we know x_2 then x_2 becomes the known approximation for the root and x_3 is the better approximation. This means that the formula for x_3 is

$$x_3 = x_2 - \frac{f(x_2)}{f'(x_2)}.$$

We see that there is a pattern here. Given one guess the formula for the next guess is always the same. If we have our nth guess x_n, then the formula for the next guess x_{n+1} is

$$x_{n+1} = x_n - \frac{f(x_n)}{f'(x_n)}.$$

Using this iterative formula for approximating the solution of the equation $f(x) = 0$ is known as using Newton's method for approximating the solution.

So far we have three guesses for the solution of $x^3 - 2x^2 - 7x + 8 = 0$. The guesses are $x_1 = 3$, $x_2 = 3.5$, and $x_3 = 3.3810$. Using our iterative formula with $n = 3$, the formula for the fourth guess is

$$x_4 = x_3 - \frac{f(x_3)}{f'(x_3)} = x_3 - \frac{x_3^3 - 2x_3^2 - 7x_3 + 8}{3x_3^2 - 4x_3 - 7}.$$

Substituting $x_3 = 3.3810$, we get

$$x_4 = 3.3810 - \frac{(3.3810)^3 - 2(3.3810)^2 - 7(3.3810) + 8}{3(3.3810)^2 - 4(3.3810) - 7}$$

$$= 3.372326229$$

Since $f(3.372326229) = 0.00061198$ and $f(3.3810) = 0.1194$, it is clear that $x = 3.372326229$ is closer to the actual solution of $x^3 - 2x^2 - 7x + 8 = 0$ than is $x = 3.3810$.

We can continue this iteration as far as we like. With each new calculation we get a value of x that is nearer and nearer the actual solution. If we let $n = 4$ in the general iteration formula, we get the formula for the fifth guess

$$x_5 = x_4 - \frac{x_4^3 - 2x_4^2 - 7x_4 + 8}{3x_4^2 - 4x_4 - 7}.$$

Substituting $x_4 = 3.372326229$ gives

$$x_5 = 3.372281324.$$

Note that $f(3.372281324) = 0.00000001 = 1.6(10^{-8})$. This indicates that x_5 is correct to about seven decimal places. If we want to determine "for

how many decimal places is this value of the solution correct", we can just find our next guess. A short computation gives

$$x_6 = 3.372281323$$

Note that $|x_5 - x_6| = 0.000000001$. We have found the value of the solution correct to at least eight decimal places.

Example 2. Find the solution of the equation $2 \sin x = x$ such that $x > 0$.

Solution. Let us sketch a graph of $y = 2 \sin x$ and $y = x$ on the same axis.

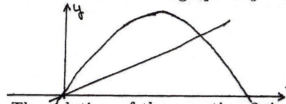

The solution of the equation $2 \sin x = x$ is the value of x where the graph of $y = 2 \sin x$ crosses the graph of $y = x$. This value of x is some number between 1 and 3. Let us check some values of $f(x) = 2 \sin x - x$.

$$f(1) = 0.68 \qquad f(2) = -0.18.$$

Clearly the equation $2 \sin x - x = 0$ has a solution between $x = 1$ and $x = 2$. The value $x = 2$ is closer to the solution. Let us use $x_1 = 2$ as our first guess. The Newton's method general formula is

$$x_{n+1} = x_n - \frac{f(x_n)}{f'(x_n)}.$$

Using $f(x) = 2 \sin x - x$ and $f'(x) = 2 \cos x - 1$, the general formula for this equation is

$$x_{n+1} = x_n - \frac{2 \sin x_n - x_n}{2 \cos x_n - 1}.$$

Let $n = 1$, we get

$$x_2 = x_1 - \frac{2 \sin x_1 - x_1}{2 \cos x_1 - 1}.$$

Substituting $x_1 = 2$ gives

$$x_2 = 2 - \frac{2 \sin 2 - 2}{2 \cos 2 - 1} = 1.900995594$$

Let $n = 2$ in the general formula, we get

$$x_3 = x_2 - \frac{2\sin x_2 - x_2}{2\cos x_2 - 1}.$$

Substituting $x_2 = 1.900995594$, we get

$$x_3 = 1.895511645.$$

The next two iterations are

$$x_4 = 1.895494267$$
$$x_5 = 1.895494267$$

Since x_4 equals x_5 for nine decimal places it follows that $x = 1.8954944267$ is the correct solution for 9 decimal places.

We would like to be able to prove that when using Newton's method each approximation to the solution is better than the previous approximation, that is

$$|f(x_{n+1})| < |f(x_n)|.$$

However, it is not possible to prove that this is true in all circumstances. In fact it is not even true in all circumstances. However, it is true in almost all circumstances where our first guess is near a solution of $f(x) = 0$. The key to using Newton's method for solving an equation is to start with a guess that is somewhat near the actual solution. Starting with a guess that is **not** near an actual solution will sometimes yield good results and sometimes it will not. Newton's method will fail, for example, if the linear approximation is $L_c(x) = $ constant. This happens when $f'(c) = 0$. In this case we cannot solve $L_c(x) = 0$ for x and find our next guess because $L_c(x) = 0$ has no solution. This would not happen if our first guess is near a solution.

Newton's method requires that you make a good first guess for the root. In particular, it requires that $f'(x)$ can not equal zero for any value of x between your guess and the actual root you are trying to approximate. In Example 2, $f'(x) = 2\cos(x) - 1$ and so $f'(\pi/3) = 0$ and $f'(5\pi/3) = 0$. This means that our first guess must be between $\pi/3 = 1.04$ and $5\pi/3 = 5.20$ for best results. A random guess will yield random results.

The usual method for finding a first guess for a root of $f(x) = 0$ is to find numbers b_1 and b_2 such that $f(b_1)$ and $f(b_2)$ have different signs. This means that there must be at least one root of $f(x) = 0$ between b_1 and b_2. You then hope that there is only one such root and that it is the root you want to find. If $f'(x)$ does not change signs for $b_1 \leq x \leq b_2$, then there is only one root of $f(x) = 0$ between b_1 and b_2 and Newton's method will find this root.

Exercises

1. Show that the equation $x^3 - 7x^2 + 25x - 35 = 0$ has a solution between $x = 2$ and $x = 3$. Find this solution correct to six decimal places.

2. Show that the equation $x^3 = 10$ has a solution between 2 and 3. Find this solution correct to eight decimal places.

3. Find all the solutions of the equation $x^3 - 3x^2 - 5x + 9 = 0$, correct to 8 decimal places.

4. Find the solution of $x^2 = 2 \sin x$ such that $x > 0$, correct to at least 6 decimal places.

5. Find the solution of $x \sin x + \cos x = 0$ such that $0 < x < \pi$.

6379 Introduction to Antiderivatives

Suppose $f(t)$ is the amount of some quantity at time t, then the rate of change of this quantity with respect to time is the derivative $f'(t)$. If $f(t)$ denotes the position of an object on a line, then $f'(t)$ is the velocity. We have considered such problems in our previous work. Now let us consider the reverse problem. Suppose we know the rate of change of a quantity and wish to find the amount. Suppose we know the velocity of an object along a line and wish to find the position. These are problems for which we would like to find a function $F(x)$ whose derivative is a known given function $f(x)$. We have studied the problem "given a function find its derivative". We now consider the inverse problem "given the derivative find the function".

Definition 1. A function $F(x)$ is called an **antiderivative** of $f(x)$ if $F'(x) = f(x)$ for all x for which $f(x)$ is defined.

Example 1. Find an antiderivative of $f(x) = x^3$.

Solution. We need to find a function $F(x)$ such that $F'(x) = x^3$. We use trial and error. First, note that $\frac{d}{dx}(x^4) = 4x^3$. Dividing both sides by 4, we get $(1/4)\frac{d}{dx}(x^4) = x^3$, or $\frac{d}{dx}[(1/4)x^4] = x^3$. Thus, if $F(x) = (1/4)x^3$, then $F'(x) = x^3$. The antiderivative of x^3 is $(1/4)x^4$. However, there are other functions that have x^3 as their derivative.

$$\frac{d}{dx}[(1/4)x^4 + 5] = x^3 \text{ and } \frac{d}{dx}[(1/4)x^4 - 20] = x^3.$$

The antiderivative is not unique. In fact it is clear that $\frac{d}{dx}[(1/4)x^4 + C] = x^3$, whenever C is any constant. Therefore

$$F(x) = (1/4)x^3 + C,$$

where C is any constant, is an antiderivative of $f(x) = x^3$. In fact this idea that we can add a constant is true for any antiderivative. Suppose $F(x)$ is an antiderivative of $f(x)$, that is, $F'(x) = f(x)$. It follows that

$$\frac{d}{dx}[F(x) + C] = F'(x) = f(x).$$

If $F(x)$ is an antiderivative of $f(x)$, then $F(x) + C$, C any constant, is also an antiderivative.

Definition 2. The collection of all antiderivatives of $f(x)$ is called the **general antiderivative** of $f(x)$.

Note that the general antiderivative is a whole collection of functions. You get a different function when you use different numbers for C. We do not like to get hung up on the fact that the general antiderivative is a collection of functions. When referring to the general antiderivative we usually refer to it as though it is a single function. When we are asked to find the antiderivative of a function, it is understood that we are expected to find the general antiderivative.

Example 2. Find the general antiderivative of $f(x) = \dfrac{1}{1+x^2}$.

Solution. Searching through our differentiation formulas until we find that

$$\frac{d}{dx}[\arctan x] = \frac{1}{1+x^2}.$$

This tells us that the general antiderivative of $f(x) = \frac{1}{1+x^2}$ is

$$F(x) = \arctan x + C.$$

Example 3. Find the general antiderivative for $f(x) = x^2 + \sin x$.

Solution. After searching our differentiation formulas we find

$$\frac{d}{dx}[(1/3)x^3] = x^2 \text{ and } \frac{d}{dx}[-\cos x] = \sin x$$

This means that

$$\frac{d}{dx}[(1/3)x^3 - \cos x] = x^2 + \sin x.$$

The general antiderivative is

$$F(x) = (1/3)x^3 - \cos x + C.$$

We need some notation to indicate antiderivatives. We use the symbol $\int$ which is called an integral. The notation

$$\int f(x)dx$$

means to find the general antiderivative of $f(x)$ or is a notation for the general antiderivative. When the notation $\int f(x)dx$ is used the function $f(x)$ is called the **integrand**. Reversing all our derivative formulas we can write down some antiderivative formulas. These may also be called integral formulas or integration formulas.

Basic Table of Antiderivatives

$$\int kx^{k-1} = x^k + C \text{ for } k \neq 0 \qquad \int x^{-1}dx = \ln x + C \qquad \int e^x dx = e^x + C$$

$$\int \cos x dx = \sin x + C \qquad \int \sin x dx = -\cos x + C \qquad \int \sec^2 x dx = \tan x + C$$

$$\int \frac{1}{1+x^2}dx = \arctan x + C \qquad \int \frac{1}{\sqrt{1-x^2}}dx = \arcsin x + C$$

$$\int \sec x \tan x dx = \sec x + C$$

We rewrite the first formula letting $k = n+1$ as

$$\int (n+1)x^n dx = x^{n+1} + C \text{ or } \int x^n dx = \frac{1}{n+1}x^{n+1} + C \text{ for } n+1 \neq 0.$$

When presented with the problem of finding an antiderivative one can rely on inverting the differentiation formulas which are already known. However, it is better to memorize the above list of antidifferentiation formulas.

We often make use of the following facts about antiderivatives. The truth of these formulas follows directly from the corresponding formulas for differentiation.

$$\int cf(x)dx = c\int f(x)dx \text{ and } \int [f(x) + g(x)]dx = \int f(x)dx + \int g(x)dx.$$

As long as we are applying the antidifferentiation formulas directly everything is easy. However, even a slight change in the function for which we need to find the antiderivative can result in a need to think in more detail before we can find the correct antiderivative. We easily find that $\int \cos x dx = \sin x + C$ as it is a direct application of the antiderivative formula. However, the problem $\int \cos(5x) dx$ does not fit directly into the formula. Our first guess might be

$$\int \cos(5x) dx = \sin(5x) + C \text{ but } \frac{d}{dx}[\sin(5x) + C] = 5\cos(5x).$$

This is not the correct function for the antiderivative. If we divide both sides by 5, we see

$$\frac{d}{dx}[(1/5)\sin(5x) + C] = \cos(5x).$$

This says that the general antiderivative is

$$\int \cos(5x) dx = (1/5)\sin(5x) + C.$$

Example 4. Find the general antiderivative (find the indefinite integral)

$$\int \frac{1}{1 + 9x^2} dx.$$

Solution. We see that this problem has $(3x)^2$ where the formula for arctan x has x^2. Our first guess for the antiderivative might be

$$\int \frac{1}{1 + 9x^2} dx = \arctan(3x) + C.$$

But using the chain rule, we find

$$\frac{d}{dx}[\arctan(3x)] = \frac{3}{1 + 9x^2}.$$

Our first guess is not quite correct. Divide both sides of this equation by 3 and we get

$$\frac{d}{dx}[(\frac{1}{3})[\arctan(3x)] = \frac{1}{1 + 9x^2}.$$

Therefore, the correct expression is

$$\int \frac{1}{1 + 9x^2} dx = (1/3)\arctan(3x) + C.$$

Exercises

Find each of the following general antiderivatives or integrals.

1. $\int x^{10} dx$

2. $\int \frac{1}{\sqrt{1 + x^2}} dx$

3. $\int (6x^2 + 5\sin x + \frac{1}{1 + x^2}) dx$

4. $\int \cos(5x) dx$

5. $\int e^{10x} dx$

6. $\int [2\sin(3x) + 5x^{-1}] dx$

7. $\int \frac{2x^4 - 3x + 5}{x^2} dx$

8. $\int (x + 5)^4 dx$

9. $\int (3x + 5)^4 dx$

10. $\int \frac{1}{1 + 16x^2} dx$

11. $\int \frac{1}{\sqrt{1 - 4x^2}} dx$

12. $\int (10x^{3/2} + 8x^{1/3}) dx$

Verify the formulas

13. $\int (7x + 2)^3 dx = \frac{1}{28}(7x + 2)^4 + C$

14. $\int (3x + 5)^{-2} dx = -\frac{1}{3}(3x + 5)^{-1} + C$

6381 Reversing the Chain Rule

The notation $F'(x) = f(x)$ means the same thing as $F'(u) = f(u)$. For example, let $F(x) = x \sin x$, then $F'(x) = x \cos x + \sin x$. The formula

$$\frac{d}{dx}[x \sin x] = x \cos x + \sin x$$

is the same as the formula

$$\frac{d}{du}[u \sin u] = u \cos u + \sin u.$$

There is only a change in the letter used as the independent variable. The formula $F'(x) = f(x)$ is the same as the formula $F'(u) = f(u)$ except for a change in the letter used as the independent variable.

Example 1. Find the antiderivative $\int \cos(10x)dx$.

Solution. We begin by using guess and check. We recall that the sine function has the cosine as its derivative,

$$\frac{d}{dx}[\sin(10x)] = 10 \cos(10x).$$

The antiderivative form of this equation is

$$\int 10 \cos(10x)dx = \sin 10x$$

This is not exactly the equation we want, but after dividing both sides by 10, we get the desired result:

$$\int \cos(10x)dx = (1/10) \sin 10x + C.$$

Example 2. Find the antiderivative $\displaystyle\int \frac{1}{1 + 25x^2}dx.$

Solution. The expression $(1 + 25x^2)^{-1}$ looks a lot like the derivative of an arctan function. Using the chain rule

$$\frac{d}{dx}[\arctan[u(x)]] = \frac{u'(x)}{1 + [u(x)]^2}.$$

This indicates we should let $u(x) = 5x$.

$$\frac{d}{dx}[\arctan(5x)] = \frac{5}{1 + (5x)^2}.$$

The antiderivative form of this is

$$\int \frac{5}{1 + 25x^2} dx = \arctan(5x).$$

This is not exactly what we want, but dividing both sides by 5 gives the desired result

$$\int \frac{1}{1 + 25x^2} dx = (1/5)\arctan(5x) + C.$$

Let us do the same solution of this problem in a slightly different format. Let $u = 5x$, then $du = (5)dx$ or $dx = (1/5)du$. Substituting gives

$$\int \frac{1}{1 + u^2} \frac{du}{5} = \frac{1}{5} \int \frac{1}{1 + u^2} du = \frac{1}{5} \arctan(u) + C$$

Replacing u with $5x$, we have

$$\int \frac{1}{1 + 25x^2} dx = \frac{1}{5} \arctan(5x) + C.$$

Example 3. Find the general antiderivative $\int \dfrac{dx}{3x + 5}$.

Solution. This antiderivative looks somewhat like $\int \frac{du}{u}$ which we can evaluate since $\frac{d}{du}[\ln u] = \frac{1}{u}$. Let us make the substitution $u = 3x + 5$ and $du = 3dx$. This converts the antiderivative to

$$\int \frac{(1/3)du}{u} = \frac{1}{3} \int \frac{du}{u} = \frac{1}{3} \ln|u| + C.$$

Replace u with $3x + 5$ and we get

$$\int \frac{dx}{3x + 5} = \frac{1}{3} \ln|3x + 5| + C.$$

Note that we are assuming that $3x + 5 \neq 0$.

This is perhaps a good time to discuss the expression $f(x) = \ln|x|$ for $x \neq 0$. Let us now consider another situation. First, suppose $x > 0$, and $f(x) = \ln x$, then we found the usual differential formula which is $f'(x) = 1/x$. Next, suppose $x < 0$, then function $g(x) = \ln(-x)$ is defined for $x < 0$. Using the chain rule

$$g'(x) = \frac{d}{dx}[\ln(-x)] = \frac{1}{-x}(-1) = \frac{1}{x}.$$

Corresponding to these two derivative formulas we really have two integral formulas

$$\int \frac{1}{x} dx = \ln(x) + C \text{ for } x > 0.$$

$$\int \frac{1}{x} dx = \ln(-x) + C \text{ for } x < 0.$$

Recall that $|x| = x$ for $x > 0$ and $|x| = -x$ for $x < 0$. These two antiderivative formulas can be combined into one formula as follows:

$$\int \frac{1}{x} dx = \ln|x| + C \text{ for } x \neq 0.$$

In the general notation for the antiderivative of $f(x)$ is

$$\int f(x) dx.$$

In this notation the dx indicates the variable of integration is x. For example

$$\int 4x^3 dx = x^4 + C \text{ while } \int 4x^3 dy = 4x^3 y + C.$$

The most important thing about the notation dx is that it behaves like a differential. We use the differential notation as part of the antiderivative notation because it behaves like a differential.

Example 4. Find the general antiderivative

$$\int (\sin^2 x)(\cos x) dx.$$

Solution. Which of our antiderivative formulas would seem to fit this problem? Notice that we have $[\sin x]^2$. Maybe we need the power formula $\int u^n du = \frac{1}{n+1} u^{n+1} + C$. Also note that $\frac{d}{dx}(\sin x) = \cos x$. After considering these facts it seems logical to try the substitution $u(x) = \sin x$. This means $du = (\cos x)dx$. Substituting this into the integral, we have

$$\int u^2 du = \frac{u^3}{3} + C.$$

Substituting back $\sin x$ for u, we have

$$\int (\sin x)^2 (\cos x)dx = (1/3)(\sin x)^3 + C.$$

Exercises

Find each of the following antiderivatives.

1. $\int e^{5x} dx$ 2. $\int \sin(5x)dx$ 3. $\int \sec^2(5x)dx$

4. $\int e^{\sin x}(\cos x)dx$ 5. $\int (\sin x)^3(\cos x)dx$ 6. $\int \frac{\cos x}{\sin x}dx$

7. $\int e^{(x^2+3x)}(2x+3)dx$ 8. $\int [\sin(x^2+3x)](2x+3)dx$

9. $\int \frac{2x+3}{x^2+3x}dx$ 10. $\int \sec(3x)\tan(3x)dx$

11. $\int e^{cx} dx$ 12. $\int \sin(cx)dx$

6383 The Substitution Rule

Integration is the reverse of differentiation. For every differentiation formula there is a companion integration formula. Suppose $F'(x) = f(x)$. The chain rule is

$$\frac{d}{dx}[F(u(x))] = F'(u(x))u'(x) = f(u(x))u'(x).$$

The companion integration rule for the chain rule is

$$\int F'(u(x))u'(x)dx = F(u(x)) + C.$$

This is usually written as follows. It is known as
The Substitution Rule

$$\int f(u(x))u'(x)dx = F(u(x)) + C \text{ where } F'(u) = f(u).$$

Finding the antiderivative is also known as finding the indefinite integral. If we suppress the x, then the substitution rule can be written as

$$\int f(u)du = F(u) + C \text{ where } F'(u) = f(u).$$

We usually use the letter u in this substitution. For this reason this is often called u substitution. When we compare this last form with the next to last form of the substitution rule we see that an important fact is true. It is permissible to operate with dx and du after the integral sign as if they were differentials. This is the reason the notation dx is used in integrals. It is clear that some part of the notation for the integral must indicate the variable of integration in case there are arbitrary constants in the function to be integrated. We use the differential dx to denote the variable of integration. At first it would seem that this is not a wise idea. We are adopting a notation that has already been assigned a meaning. We are giving the symbol dx a new job which is to indicate the variable of integration in $\int f(x)dx$. We use the notation dx instead of say (x) because the dx in this situation can be treated as a differential. It obeys the usual

249

rules for differentials. In future we will assume that rules about differentials apply to dx when written in $\int f(x)dx$.

Basically we classify "find the antiderivative" problems into two kinds. The first kind is "this problem is fairly easy" like the first couple of antiderivative problems we discussed. The second kind, the harder kind, is the kind for which "we must make use of the companion rule to the chain rule, the Substitution Rule". In most problems where we are asked to find the antiderivative the Substitution Rule is used to convert the problem to a problem where we need only use one of the formulas in the Basic Table of Antiderivatives. Another good reason for memorizing the Basic Table of Antiderivatives. In this discussion two of the formulas in the Basic Table of Antiderivatives are slightly special. The special formulas are

$$\int \frac{dx}{1+x^2} = \arctan(x) + C \text{ and } \int \frac{dx}{\sqrt{1-x^2}} = \arcsin(x) + C.$$

We will discuss antiderivatives involving these formulas later.

Most of the hard problems where we are asked to find the antiderivative of a function $f(x)$ in calculus class are solved using the following rule.

Rule. First find the most complicated composite function in the integrand. Second, find the first inside function for this composite function. Use the substitution $u(x)$ equal to this first inside function. Apply the Substitution Rule for finding the antiderivative.

The reason this rule works is that all complicated functions for which we are asked to find the antiderivative are the result of differentiation using the chain rule. The chain rule usually results in the product of two functions. Examine the integrand to see if it is the product of two functions. In the following example the product includes a quotient.

Example 1. Find the general antiderivative

$$\int \frac{x+5}{(x^2 + 10x + 34)^{3/2}} dx.$$

Solution. What is the most complicated composite function which is a factor of the integrand? The most complicated composite function should be multiplied times the rest of the integrand. We could rewrite this problem as

$$\int (x^2 + 10x + 34)^{-3/2}(x+5)dx.$$

However, all this heavy thinking is not necessary. Clearly the most complicated composite function is $(x^2 + 10x + 34)^{3/2}$. The first inside function is $x^2 + 10x + 34$. This means that we make the substitution

$$u = x^2 + 10x + 34$$
$$du = (2x + 10)dx = 2(x+5)dx.$$

After making this substitution the antiderivative problem becomes

$$\int u^{-3/2}\frac{du}{2} = -u^{-1/2} + C.$$

Applying the Substitution Rule, that is, replacing u with $u = x^2 + 10x + 34$, we get

$$\int \frac{x+5}{(x^2 + 10x + 34)^{3/2}}dx = -\frac{1}{(x^2 + 10x + 34)^{1/2}} + C.$$

In general finding the antiderivative $\int f(x)dx$ is a three step process. First, we look at $f(x)$ and determine the substitution $u = u(x)$. Note that $g(u(x))du(x) = f(x)dx$. Second, the result is $\int g(u)du$ where we know the antiderivative of $g(u)$, that is, $\int g(u)du = G(u) + C$. Third the substitution Rule says $\int g(u(x))du(x) = G(u(x)) + C$.

Example 2. Find the general antiderivative

$$\int [(x^3 + 6x)^{2/3}(x^2 + 2) + x\sin(3x^2 + 5)]dx.$$

Solution. It is clear that this integrand is the sum of two complicated expressions and so should be treated as a sum. In many cases where the integrand is a sum we make use of the rule

$$\int [f(x) + g(x)]dx = \int f(x)dx + \int g(x)dx.$$

251

Consider the first integrand $(x^3 + 6x)^{3/2}(x^2 + 2)$. The most complicated composite function is $(x^3 + 6x)^{3/2}$ and the first inside function is $x^3 + 6x$. Therefore, let $u(x) = x^3 + 6x$, then $du = (3x^2 + 6)dx = 3(x^2 + 2)dx$. Making this substitution in the first integral, we get

$$\int u^{2/3} \frac{du}{3} = \frac{1}{3} \frac{u^{5/3}}{5/3} + C = \frac{1}{5} u^{5/3} + C.$$

Applying the Substitution Rule we replace u with $u = x^3 + 6x$ and have

$$\int (x^3 + 6x)^{2/3}(x^2 + 2)dx = \frac{1}{5}(x^3 + 6x)^{5/3} + C.$$

The second antiderivative is

$$\int x \sin(3x^2 + 5)dx.$$

The most complicated composite function is $\sin(3x^2 + 5)$ and the first inside function is $3x^2 + 5$. Let $u(x) = 3x^2 + 5$ and so $du = (6x)dx$. Making this substitution, we get

$$\int \sin(u) \frac{du}{6} = -\frac{1}{6} \cos(u) + C.$$

Applying the Substitution Rule we replace u with $(3x^2 + 5)$ and get

$$\int x \sin(3x^2 + 5)dx = -\frac{1}{6} \cos(3x^2 + 5) + C.$$

It follows that

$$\int [(x^3 + 6x)^{2/3}(x^2 + 2) + x \sin(3x^2 + 5)]dx = \frac{1}{5}(x^3 + 6x)^{5/3} - \frac{1}{6} \cos(3x^2 + 5) + C.$$

Example 3. Find the general antiderivative (integral)

$$\int \sec^2(9x)dx.$$

Solution. The composite function is $[\sec(9x)]^2$. Following the above rule and letting $u = \sec(9x)$ does not yield a solution. We should suspect that it will not work because the integrand is not a product of $[\sec(9x)]^2$ times another function. The very first thing we notice is that we must find the antiderivative of $\sec^2$ and we know that $\int \sec^2 u\, du = \tan u$. The difficulty is that we have $9x$ where we would like to have just the simple independent variable. We can make $9x$ into a simple variable by letting $u = 9x$, then $9dx = du$ or $dx = (1/9)du$. Making this substitution, we get

$$\int (\sec^2 u)\frac{du}{9} = \frac{1}{9}\int (\sec^2 u)du = \frac{1}{9}\tan u + C.$$

Replace u with $9x$ and we get the desired result

$$\int \sec^2(9x)dx = \frac{1}{9}\tan(9x) + C.$$

In this solution, we did not follow the rule where we first find the most complicated composite function. The reason is that we are integrating secant squared and we have a formula for this antiderivative in our basic table of antiderivatives. This means we need $\int \sec^2 u\, du$ and so we must let $u = 9x$. The over riding idea in every antiderivative problem is that we are trying to make a u substitution which converts the integrand to one of the formulas in our basic table of antiderivatives. We should begin every problem by looking to see if we easily see a u substitution that converts the problem to one of the formulas in our basic table of antiderivatives.

In this problem, there is a simple formula for u, namely, $u = 9x$. When there is a simple formula for u we sometimes decide to just keep the value of u in our head and never write the expression for u on paper. Keeping the value of u in our head often leads to mistakes. Always write down the value of $u(x)$ as part of the solution of an antiderivative problem.

Example 4. Find the general antiderivative $\displaystyle\int \frac{x}{3x^2 + 5}dx.$

Solution. Suppose that we do not see at once how to fit this into one of the formulas in our basic table of antiderivatives. We then apply the Rule. At first, we may not see a composite function. However, if we rewrite the problem as

$$\int x(3x^2 + 5)^{-1}dx,$$

then the most complicated composite function is $(3x^2+5)^{-1}$ and the first inside function is $3x^2+5$. Let $u(x) = 3x^2+5$, then $du = (6x)dx$. Substituting $u = 3x^2 + 5$ into the integral, we get

$$\int \frac{(1/6)}{u} du = \frac{1}{6} \int \frac{du}{u} = (1/6)\ln u + C$$

Let $u = 3x^2 + 5$ in the Substitution Rule and we get

$$\int \frac{x}{3x^2 + 5} dx = (1/6)\ln(3x^2 + 5) + C.$$

We could have solved this problem more directly by observing that $\frac{d}{dx}(3x^2 + 5) = 6x$ and so the substitution $u = 3x^2 + 5$ converts to a problem involving $\int \frac{du}{u}$. Note that $3x^2 + 5 > 0$.

Example 5. Find the expression for the general antiderivative (integral)

$$\int [\sin(5x)]^3 \cos(5x)dx.$$

Solution. What part of this integrand is the most complicated composite function? Clearly it is $[\sin(5x)]^3$. What is the first inside function for this composite function? Clearly it is $\sin(5x)$. Let $u = \sin 5x$. This means $du = 5\cos(5x)dx$ or $\cos(5x)dx = (1/5)du$. Substituting this into the antiderivative we get

$$\int u^3(1/5)du = (1/5) \int u^3 du = \frac{1}{5}\frac{u^4}{4} + C.$$

Replacing u with $\sin(5x)$, we get

$$\int [\sin(5x)]^3 \cos(5x)dx = (1/20)[\sin(5x)]^4 + C.$$

Example 6. Find the integral (antiderivative)

$$\int \frac{\arcsin(x/3)}{\sqrt{9 - x^2}} dx.$$

254

Solution. Suppose we try to fit this into the find most complicated composite rule. If we try something simple like $u = x/3$ on $u = 9 - x^2$ we see that it does not simplify the integrand enough to yield a formula from our table of basic antiderivative formulas. It is helpful to look at the integrand $f(u(x))u'(x)$ in the substitution rule and note that it is a product. We can rewrite the problem with the integrand expressed as a product

$$\int [\arcsin(x/3)](9 - x^2)^{-1/2} dx.$$

One factor must be $f(u(x))$ and the other $u'(x)$. Choose one factor as $f(u(x))$. If we make a wrong choice just go back and start over using the other choice. Here we really need to recall the differentiation formula

$$\frac{d}{du}[\arcsin u] = \frac{1}{\sqrt{1 - u^2}}.$$

This makes it clear that we need to use the substitution $u = \arcsin(x/3)$. We have

$$du = \frac{1/3}{\sqrt{1 - (x/3)^2}} dx = \frac{1}{\sqrt{9 - x^2}} dx.$$

Making this substitution the given integral becomes

$$\int u\, du = (1/2)u^2 + C.$$

Let $u = \arcsin(x/3)$, then the Substitution Rule says

$$\int \frac{\arcsin(x/3)}{\sqrt{9 - x^2}} dx = (1/2)[\arcsin(x/3)]^2 + C \text{ for } |x| < 3.$$

Looking back at the solution of this problem we see that the most complicated composite function was $[\arcsin(x/3)]^1$. Not easy to see. It would have been more obvious had it been $[\arcsin(x/3)]^3$.

Exercises

Find each of the following antiderivatives.

1. $\int \sin(5x + 8)dx$

2. $\int e^{\cos bx} \sin bx\,dx$

3. $\int [\cos(bx)]^4 \sin(bx)dx$

4. $\int \dfrac{dx}{\sqrt{3x + 5}}$

5. $\int \dfrac{x}{8x^2 + 11}dx$

6. $\int x^{-1/2}e^{\sqrt{x}}dx$

7. $\int (x + 3)\sqrt{x^2 + 6x + 10}\ dx$

8. $\int \dfrac{(x + 4)}{(3x^2 + 24x + 20)^{8/5}}\ dx$

9. $\int [\sin(5x^2 + 8)]^3[\cos(5x^2 + 8)](x)dx$

10. $\int \dfrac{\sin(5x)}{(9 + 4\cos 5x)^{3/4}}dx$

11. $\int [5 + 2\tan(3x)]^{2/3}[\sec(3x)]^2 dx$

12. $\int [\cosh(8x)]^2 \sinh(8x)dx$

13. $\int \dfrac{\arcsin(x/4)}{\sqrt{16 - x^2}}dx$

14. $\int \dfrac{[\arctan(x/3)]^2}{9 + x^2}dx$

6385 More Substitution Rule

We are now going to discuss some "find the antiderivative" problems that are solved by fitting the integrand into one of two antiderivative formulas. These problems can not really be solved by using the rule about finding the most complicated composite function. These antiderivative formulas are

$$\int \frac{dx}{\sqrt{1-x^2}} = \arcsin(x) + C \text{ and } \int \frac{dx}{1+x^2} = \arctan(x) + C.$$

Example 1. Find the general antiderivative

$$\int \frac{dx}{25 + 36x^2}.$$

Solution. Note that we can not find this antiderivative by first finding the most complicated composite function. Instead we need to look at the antiderivative formula

$$\int \frac{du}{1+u^2} = \arctan(u) + C.$$

The integrand $\frac{1}{25+36x^2}$ looks a lot like $\frac{1}{1+u^2}$. We should be able to convert this integrand into the form $\frac{1}{1+u^2}$. Note that

$$\frac{1}{25 + 36x^2} = \frac{1}{25[1 + (36x^2)/25]}.$$

Letting $u^2 = (36x^2)/25$ should work. Let us make the substitution $u = (6x)/5$ or $x = (5u)/6$, then $dx = (5/6)du$. Making this substitution gives

$$\int \frac{(5/6)du}{25 + 36(25/36)u^2} = \frac{1}{30} \int \frac{du}{1 + u^2}$$

$$= \frac{1}{30} \arctan(u) + C.$$

Applying the Substitution Rule and replacing u with $(6x)/5$ gives

$$\int \frac{dx}{25 + 36x^2} = \frac{1}{30} \arctan(\frac{6x}{5}) + C.$$

Example 2. Find the general antiderivative

$$\int \frac{dx}{\sqrt{81 - 16x^2}}.$$

Solution. The key to this problem is to recall the basic antiderivative formula

$$\int \frac{du}{\sqrt{1 - u^2}} = \arcsin(u) + C.$$

We note that $\sqrt{81 - 16x^2}$ has the form $\sqrt{1 - u^2}$. A little algebra gives

$$\sqrt{81 - 16x^2} = \sqrt{81[1 - (16x^2)/81]} = 9\sqrt{1 - (4x/9)^2}.$$

We need to use the substitution $u = 4x/9$ or $x = 9u/4$. This gives $dx = (9/4)du$. Making this substitution, we get

$$\int \frac{(9/4)du}{\sqrt{81 - 81u^2}} = \frac{1}{4} \int \frac{du}{\sqrt{1 - u^2}} = \frac{1}{4} \arcsin(u) + C.$$

Replacing u with $(4x)/9$ and applying the Substitution Rule, we get

$$\int \frac{dx}{\sqrt{81 - 16x^2}} = \frac{1}{4} \arcsin(\frac{4x}{9}) + C.$$

In almost every find the antiderivative problem we must as we did here use the Substitution Rule. Suppose $u = u(x)$ and $F'(u) = f(u)$, then the Substitution Rule can be written as

$$\int f(u(x))u'(x)dx = F(u(x)) + C.$$

It is easy to see that the companion derivative form of this is the chain rule.

$$\frac{d}{dx}[F(u(x))] = F'(u(x))u'(x) = f(u(x))u'(x).$$

We usually use the letter u in this substitution. For this reason this is often called u-substitution.

In a somewhat general way we begin the solution of "find the antiderivative" problems in three steps.

Step 1. Study the problem and see if we can see a u substitution that converts the problem into one of the formulas in the Basic Table of Antiderivatives. If you can discover such a function $u(x)$, then make the substitution and finish the problem.

Step 2. Check to see if the problem is of the form $\int \frac{du}{1+u^2}$ or $\int \frac{du}{\sqrt{1-u^2}}$. If it is, then proceed as in the above example.

Step 3. Select the most complicated composite function and let $u(x)$ equal to the first inside function. This must work on any "calculus class" problem.

Again suppose we have the problem of finding an antiderivative. The most direct way to solve this problem is to recognize that there is a u-substitution which converts the problem into one of the antiderivative in the Basic Table of Antiderivatives. This is Step 1 above. If the antiderivative problem involves one of the two formulas illustrated in Example 1 and Example 2, then this is the only way to solve the problem. If the integrand is the product or quotient of two expressions there is a very good chance you will need to look for the most complicated composite function. In the cases when we do not see how a problem fits into one of the formulas in Basic Table of Antiderivatives right away, then we look for the most complicated composite function and try letting u equal the first inside function. Composite functions always look simpler after a u substitution. The antiderivative for a non "calculus class problem" may not exist in terms of elementary functions.

Example 3. Find the general antiderivative

$$\int \frac{x+9}{4x^2+5}.$$

Solution. The key to this problem is to begin by writing the integrand as a sum.

$$\int \left[\frac{x}{4x^2+25} + \frac{9}{4x^2+25} \right] dx = \int \frac{x}{4x^2+25} dx + \int \frac{9}{4x^2+25} dx.$$

Whenever the integrand can be written as a sum as can be done in this example, we must always consider the possibility that the first step in the solution is to write the integrand as a sum as was done here. For the first antiderivative we return to what has worked so often before, we look for the most complicated composite function. The x in the numerator means that it is not of the form $\int \frac{du}{1+u^2}$. The composite function is $(4x^2 + 25)^{-1}$. Let $u = 4x^2 + 5$ and so $du = (8x)dx$, on $xdx = (1/8)du$. Substituting this into the first integral, we get

$$\int \frac{(1/8)du}{u} = \frac{1}{8} \int \frac{du}{u} = \frac{1}{8} \ln |u| + C.$$

Applying the Substitution Rule and replacing u with $4x^2 + 25$, we get

$$\int \frac{x}{4x^2 + 25} dx = \frac{1}{8} \ln(4x^2 + 25) + C.$$

The second antiderivative $\int \frac{dx}{4x^2 + 25}$ is evaluated by noting that $\frac{1}{4x^2 + 25}$ is of the form $\frac{1}{1+u^2}$. A little algebra gives

$$4x^2 + 25 = 25[1 + (4x^2)/25]$$
$$= 25[1 + (2x/5)^2].$$

Use the substitution $u = 2x/5$ or $x = (5/2)u$ so that $dx = (5/2)du$. The second antiderivative can be written as

$$\int \frac{(5/2)du}{25(1 + u^2)} = \frac{1}{10} \int \frac{du}{1+u^2} = \frac{1}{10} \arctan(u) + C.$$

Applying the Substitution Rule and substituting $u = 2x/5$ gives

$$\int \frac{dx}{4x^2 + 25} = \frac{1}{10} \arctan(\frac{2x}{5}) + C.$$

It follows that

$$\int \frac{x + 9}{4x^2 + 25} dx = \frac{1}{8} \ln(4x^2 + 25) + \frac{9}{10} \arctan(\frac{2x}{5}) + C.$$

Example 4. Find the general antiderivative

$$\int \frac{x^2 + 4x}{4x^3 + 24x^2 + 25} dx.$$

Solution. When looking at this problem the ideal thing is to notice that $\frac{d}{dx}(4x^3 + 24x^2 + 25) = 12(x^2 + 4x)$. This enables us to see that this fits into $\int \frac{du}{u}$. But suppose we do not notice this fact. Let us rewrite the integral as $\int (4x^3 + 24x^2 + 25)^{-1}(x^2 + 4x)dx$. The most complicated composite function is $(4x^3 + 24x^2 + 25)^{-1}$ and the first inside function is $(4x^3 + 24x^2 + 25)$. Therefore, let us use the substitution $u = 4x^3 + 24x^2 + 25$. This means that $du = (12x^2 + 48x)dx = 12(x^2 + 4x)dx$. Making this substitution, we get

$$\frac{1}{12} \int \frac{du}{u} = \frac{1}{12} \ln |u| + C.$$

Replacing u with $4x^3 + 24x^2 + 25$, we have

$$\int \frac{x^2 + 4x}{4x^3 + 24x^2 + 25} dx = \frac{1}{12} \ln |4x^3 + 24x^2 + 25| + C.$$

Example 5. Find the general antiderivative

$$\int \frac{\sin 5x}{(4 - \cos 5x)^3} dx.$$

Solution. The most complicated composite function is $(4 - \cos 5x)^3$. The first inside function is $4 - \cos 5x$. Therefore, we make the substitution $u = 4 - \cos 5x$. This means that $du = 5 \sin(5x)dx$ or $[\sin(5x)]dx = (1/5)du$. After we make this substitution the integral becomes

$$\int \frac{(1/5)du}{u^3} = \frac{1}{5} \int u^{-3} du = \frac{1}{5} \frac{u^{-2}}{(-2)} + C = -\frac{1}{10} \frac{1}{u^2} + C.$$

Applying the Substitution Rule and replacing u with $4 - \cos 5x$, we get

$$\int \frac{\sin 5x}{(4 - \cos 5x)^3} = -\frac{1}{10} \frac{1}{(4 - \cos 5x)^2} + C.$$

We are able to solve this problem a little quicker if we notice right away that $\frac{d}{dx}[4-\cos(5x)] = 5\sin(5x)$. This tells us that we should let $u = 4 - \cos(5x)$.

Example 6. Find the general antiderivative

$$\int x(5x+8)^{2/3}dx.$$

Solution. Note that $\frac{d}{dx}(5x+8) \neq x$. The most complicated composite function is $(5x+8)^{2/3}$. The first inside function is $5x+8$. Therefore, we make the substitution $u = 5x+8$. Then $du = 5dx$. This means that

$$x = \frac{u-8}{5} \text{ and } dx = \frac{1}{5}du.$$

Making this substitution, we have

$$\int \frac{u-8}{5}u^{2/3}\frac{du}{5} = \frac{1}{25}\int(u^{5/3} - 8u^{2/3})du$$

$$= \frac{1}{25}\left[\frac{u^{8/3}}{8/3} - 8\frac{u^{5/3}}{5/3}\right] + C = \frac{3}{200}u^{8/3} - \frac{24}{125}u^{5/3} + C.$$

Applying the Substitution Rule and replacing u with $5x+8$, we get

$$\int x(5x+8)^{2/3}dx = \frac{3}{200}(5x+8)^{8/3} - \frac{24}{125}(5x+8)^{5/3} + C.$$

If need be we can rewrite this antiderivative as

$$[5x+8]^{5/3}\left[\frac{3}{40}x - \frac{9}{125}\right] + C.$$

Example 7. Find the general antiderivative

$$\int [\tan(5x^2)]^3[\sec(5x^2)]^2(x)dx.$$

Solution. Although it is not completely clear the most complicated composite function is $[\tan(5x^2)]^3$. The integrand is the product of $[\tan(5x^2)]^3$

and $[\sec(5x^2)]^2$ so one of these must be the most complicated composite function. The absolute clue that $[\tan(5x^2)]^3$ is the most complicated composite function is the fact that the derivative of $\tan u$ is $\sec^2 u$. The first inside function is $\tan(5x^2)$. Let $u = \tan(5x^2)$, then $du = 10x[\sec(5x^2)]^2 dx$. This says that $[\sec(5x^2)]^2 (x)dx = (1/10)du$. Making this substitution, we get

$$\frac{1}{10} \int u^3 du = \frac{1}{40}u^4 + C.$$

Replacing u with $\tan(5x^2)$, we have

$$\int [\tan(5x^2)]^3 [\sec(5x^2)]^2 (x)dx = \frac{1}{40}[\tan(5x^2)]^4 + C.$$

None of our integration methods would find

$$\int \frac{\sin(3x)dx}{\sqrt{4x + \cos 5x}}$$

because the integrand is not the derivative of any composite function composed of elementary functions. It is not possible to express this antiderivative in terms of our elementary functions. You might try this antiderivative in your calculator.

Exercises

Find the following general antiderivatives

1. $\int \frac{x(x^2+3)}{5x^4+30x^2+18} dx$

2. $\int \frac{\cos(4x)}{5+3\sin(4x)} dx$

3. $\int [\tan(x^2 + 10)]^3 [\sec(x^2 + 10)]^2 (x)dx$

4. $\int \frac{x}{\sqrt{2x+5}} dx$

5. $\int x\sqrt{5x + 2}dx$

6. $\int \frac{\sin 5x}{\sqrt{4+2\cos 5x}} dx$

7. $\int \frac{dx}{\sqrt{100-81x^2}}$

8. $\int \frac{dx}{16+25x^2} dx$

9. $\int \frac{36x+25}{9x^2+64} dx$

10. $\int \frac{8x^2+x+68}{4x^2+25} dx$.

6387 The Area Problem

We are going to consider the problem of finding the area of a region in the plane. The first question is: what do we mean by the phrase "find the area"? The fundamental idea of area is that the area of a rectangle is equal to the length times the width. From this fact it follows that the area of a triangle is one half the base times the height. When we say that area is additive we mean the following. Suppose a larger region is divided into two smaller regions so that the only points the two smaller regions have in common is their common boundary. When this is the case, then the sum of the areas of the two smaller regions is equal to the area of the larger region. The fact that this is true is expressed by saying "area is additive". Using the fact that area is additive we can find the area of any region whose boundary is composed of line segments. The area of a region whose sides are line segments is found by dividing it into triangles and then adding the areas of the triangles. It is not easy to find the area of a region with some curved sides. We can not divide such a region into triangles. We all have a general idea of what we mean by area, but when we consider a region with some curved sides we need to have a more exact idea of what we mean by "find the area". We will not try here to do a logical development of the idea of area from axioms. We will just depend upon everyone's "common knowledge" of the properties of area. The most important part of everyone's common knowledge is that area is additive. Another part of everyone's common knowledge is that if region B is entirely inside region D, then the area of B is less than the area of D.

Suppose we have a region that has a single section of its boundary that is not a line segment. We are going to develop a method for finding the area of such a region. As the first step in solving this problem we consider a particular region. In order to have a clear situation we assume we have a region R in the xy plane. We assume the region is bounded by the two vertical lines $x = a$ and $x = b$ with $a < b$. A third side of the region is the x axis whose equation is $y = 0$. The fourth side of the region is a curve which is the graph of a continuous function $f(x)$. This may not seem like a very general region, but once we can find the area of regions such as this one we can generalize to any region with a curvilinear boundary.

Our first step in finding the area of this region is to divide the interval $a \leq x \leq b$ on the x axis into several subintervals of equal length. Let n

denote the number of such subintervals and Δx denote the length of each of the subintervals. It is clear that

$$\Delta x = \frac{b-a}{n}.$$

Denote the division points on the x axis by $a = x_0, x_1, x_2, \ldots x_n = b$, then the formula for the kth division point is

$$x_k = a + k(b-a)/n \text{ where } k = 0, 1, 2, \ldots n.$$

In order to get a clearer understanding of what we are doing, let us look at the situation in the case where $a = 1$, $b = 3$, and $n = 4$. The length of the subintervals in this case is given by $\Delta x = (3-1)/4 = 0.5$. The division points on the x axis are

$$x_0 = 1 \quad x_1 = 1.5 \quad x_2 = 2 \quad x_3 = 2.5 \quad x_4 = 3$$

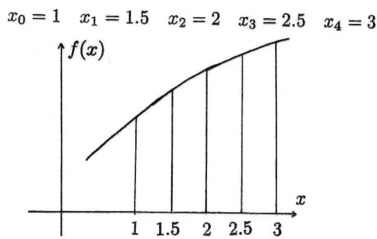

Draw vertical lines for each of these values of x. It is clear that we have now divided the larger region into four smaller regions. If we know the areas of the four smaller regions, then we can add these numbers together and get the area of the larger region. Let us concentrate on finding the area of the first small region. We need to find the area of this small region, but it has one curved side. We only know how to find the area of a rectangle. Let us approximate the area of this small region with the area of a long thin rectangle. The short side of each small rectangle is the subinterval on the x axis. How tall should the rectangle be? Let us take the value of the function $f(x)$ in the middle of the subinterval as the height of the rectangle. The midpoint of the first subinterval $1 \le x \le 1.5$ is at $x = 1.25$. Therefore, we use $f(1.25)$ as the height of this rectangle. Let us look at the graph of the curve and this first small rectangle.

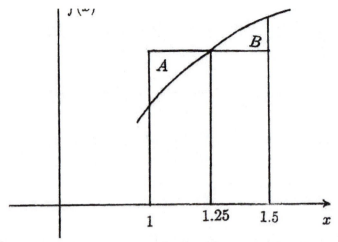

The area enclosed by the rectangle is not exactly the area under the curve. There is a small region enclosed by the rectangle which is not under the curve. This region is labeled A. There is a small region which is not enclosed by the rectangle but which is under the curve. This region's labeled B. The area of the rectangle equals the area under the curve plus area A minus area B. Since these small areas are about the same size, that is, area A is approximately equal to area B, it seems reasonable to say that area A minus area B is approximately equal to zero. This says that the area enclosed by the rectangle is approximately equal to the area under the curve between the lines $x = 1$ and $x = 1.5$. The area of this rectangle is $(0.5)f(1.25)$.

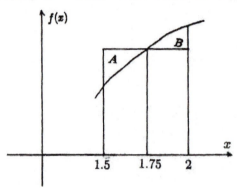

Consider the second subinterval $1.5 \leq x \leq 2$. The midpoint of this subinterval is $x = 1.75$. Use the second subinterval as the short side of a rectangle. Make the height of this rectangle $f(1.75)$. The area is $(0.5)f(1.75)$. Using a discussion similar to the one we used for the first rectangle we conclude that the area of this rectangle is approximately equal to the area under the curve between the lines $x = 1.5$ and $x = 2$.

Consider the third subinterval $2 \leq x \leq 2.5$. The midpoint of this subinterval is $x = 2.25$. Use the third subinterval as the short side of a rectangle. Make the height of this rectangle $f(2.25)$. The area is $(0.5)f(2.25)$. Using

266

a discussion similar to the one we used for the first rectangle we conclude that the area of this rectangle is approximately equal to the area under the curve between the lines $x = 2$ and $x = 2.5$.

Consider the fourth subinterval $2.5 \leq x \leq 3$. The midpoint of this subinterval is $x = 2.75$. Use the fourth subinterval as the short side of the rectangle. Make the height of this rectangle $f(2.75)$. The area is $(0.5)f(2.75)$. Using a discussion similar to the one used for the first rectangle we conclude that the area of this rectangle is approximately equal to the area under the curve between the lines $x = 2.5$ and $x = 3$.

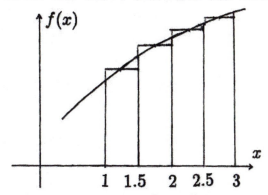

In conclusion, we say that the area under the curve $y = f(x)$ between the lines $x = 1$ and $x = 3$ is approximately equal to the sum of the areas of these four rectangles. The area of these four rectangles taken together is

$$(0.5)[f(5/4) + f(7/4) + f(9/4) + f(11/4)].$$

Let us do another approximation to the area of this same region under this same curve. Suppose that instead of dividing the interval $1 \leq x \leq 3$ into four subintervals we divide it into eight subintervals. In this case the length of each subinterval is $1/4$. This time we have eight small rectangles. For this computation we have two rectangles where we had one before. The short side of each rectangle is the subinterval on the x axis of length $1/4$. Again we take as the height of each small rectangle the value of the function $f(x)$ evaluated at the midpoint of the short subinterval. The midpoint of the first subinterval is $x = 9/8$. The area of the first long thin rectangle in this subdivision is (width) times (length) equal $(1/4)f(9/8)$. The sum of the area of all eight small rectangles is

$$(1/4)[f(9/8) + f(11/8) + f(13/8) + \ldots + f(21/8) + f(23/8)].$$

Using the same discussions as the ones used above it is clear that the sum of the areas of the eight rectangles is a good approximation to the area under the curve $y = f(x)$ between the lines $x = 1$ and $x = 3$. It is fairly clear geometrically that when we approximate the area under the curve using eight rectangles we get a better approximation to the area than we got when we used four rectangles. In general, the more rectangles we use the better the approximation. Let us keep in mind during the remainder of our discussion, the fact "the more rectangles the better the approximation".

Now that we have some idea of how this method for approximating the area of a region under a curve works, let us return to the general case. Suppose we have a region, R, bounded by the lines $x = a$, $x = b$ with $a < b$, the x axis, and the curve which is the graph of $y = f(x)$. We assume that $f(x) > 0$ and $f(x)$ is continuous. Recall that we divide the interval $a \leq x \leq b$ into n subintervals of equal length. The length is $\Delta x = (b-a)/n$. The division points marking the end points of these short subintervals are given by

$$x_k = a + k(b - a)/n \text{ with } k = 0, 1, 2, \ldots n.$$

The kth subinterval is

$$a + \frac{b - a)}{n}(k - 1) \leq x \leq a + \frac{b - a}{n}(k) \text{ with } k = 1, 2, \ldots n.$$

The midpoint of the kth subinterval is found by taking the average of the two end points. The midpoint is

$$\frac{1}{2}[a + \frac{(b - a)}{n}(k-1) + a + \frac{b - a}{n}(k)] = a + (2k - 1)\frac{b - a}{2n} \text{ where } k = 1, 2, \ldots n.$$

The width of each small rectangle is the length of the subinterval and is $\Delta x = (b - a)/n$. The height of the kth long thin rectangle is given by the value of the function evaluated at the midpoint of the kth small subinterval. The height is

$$f\left(a + \frac{2k - 1}{2n}(b - a)\right).$$

The area of the kth long thin rectangle is width times height, that is,

$$\frac{b - a}{n}f\left(a + \frac{2k - 1}{2n}(b - a)\right).$$

We add up the area of all n of the small rectangles. The sum is

$$\frac{b-a}{n} \sum_{k=1}^{n} f\left(a + \frac{2k-1}{2n}(b-a)\right).$$

This is a good approximation to the area under the curve. Also the larger the value of n the better the approximation. This says that as n approaches infinity the value of this sum must approach the actual area. This causes us to make the following definition.

Definition. Suppose $f(x)$ is continuous and $f(x) > 0$ for $a \leq x \leq b$. The area of the region under the curve $y = f(x)$ and between the lines $x = a$ and $x = b$ is given by the following limit.

$$\text{Area} = \lim_{n \to \infty} \frac{b-a}{n} \sum_{k=1}^{n} f\left(a + \frac{2k-1}{2n}(b-a)\right).$$

Example 1. Consider the region bounded by the lines $x = 2$ and $x = 5$, the x axis, and the curve which is the graph of $y = 4x^2 + 5x$. Divide this interval into $n = 4$ subdivisions. Approximate this area using these four rectangles.

Solution. The length of each subinterval is $(5 - 2)/4 = 3/4$. The formula for the division points on the x axis is

$$x_k = a + k\frac{b-a}{n} = 2 + k\frac{3}{4}, \quad k = 0, 1, 2, 3, 4.$$

The division points for the subintervals on the x axis are

$$x_0 = 2, \ x_1 = 11/4, \ x_2 = 14/4, \ x_3 = 17/4, \ x_4 = 5.$$

The midpoint of the first subinterval is $(1/2)[2 + (11/4)] = 19/8$. The midpoint of the second interval is $(1/2)[(11/4) + (14/4)] = 25/8$. The midpoints of the other two intervals are $31/8$ and $37/8$. We find the heights of these rectangles by evaluating the function at these midpoints. The heights of the long thin rectangles are

$$f(19/8) = 4(19/8)^2 + 5(19/8) = 551/16$$

$$f(25/8) = 4(25/8)^2 + 5(25/8) = 875/16$$

$$f(31/8) = 4(31/8)^2 + 5(31/8) = 1271/16.$$

$$f(37/8) = 4(37/8)^2 + 5(37/8) = 1739/16.$$

The length of the short side of each rectangle is $3/4$. The area of each rectangle is $3/4$ times the height. The sum of the areas of the four small rectangles is

$$(3/4)[(551/16) + (875/16) + (1271/16) + (1739/16)]$$

$$= (3/4)(4436/16) = 3327/16.$$

This is an approximate value of the area under the curve.

Example 2. Find the area of the region under the curve $y = 3x^2 + 4$ which is also bounded by the lines $x = 0$, $x = 4$, and the x axis.

Solution. We have $a = 0$, $b = 4$, $b - a = 4$, and $f(x) = 3x^2 + 4$. We divide the interval $0 \le x \le 4$ into n subintervals each of length $4/n$. The division points for the subintervals on the x axis are $x_k = a + \frac{b-a}{n}k = \frac{4k}{n}$, $k = 0, 1, 2, \ldots n$. The midpoint of the kth subinterval is

$$a + \frac{2k - 1}{2n}(b - a) = \frac{2k - 1}{2n}(4) = \frac{2(2k - 1)}{n}.$$

The value of the function at the midpoint of the kth subinterval is

$$f\left(\frac{4k - 2}{n}\right) = 3\left(\frac{4k - 2}{n}\right)^2 + 4 = \frac{12}{n^2}(2k - 1)^2 + 4.$$

This is the height of the kth long thin rectangle. The width of each rectangle is $4/n$. The area of the kth rectangle (length times width) is

$$\frac{4}{n}\left[\frac{12}{n^2}(2k - 1)^2 + 4\right] = \frac{48}{n^3}(2k - 1)^2 + \frac{16}{n}.$$

The sum of the areas of all rectangles is

$$\sum_{k=1}^{n}\left[\frac{48}{n^3}(2k - 1)^2 + \frac{16}{n}\right] = \frac{48}{n^3}\sum_{k=1}^{n}(2k - 1)^2 + \frac{16}{n}\sum_{k=1}^{n}(1).$$

Using the fact that $\sum_{k=1}^{n}(1) = n$ and $\sum_{k=1}^{n}(2k-1)^2 = (1/3)n(4n^2-1)$, this is equal to

$$\frac{48}{n^3}\left[\frac{n}{3}(4n^2-1)\right] + \frac{16}{n}(n) = \frac{4}{3}(48) - \frac{16}{n^2} + 16 = 80 - \frac{16}{n^2}.$$

Taking the limit as n approaches infinity, we easily find

$$\text{Area} = \lim_{n\to\infty}\left(80 - \frac{16}{n^2}\right) = 80.$$

Exercises

1. Suppose $x_k = a + k(b-a)/n$ for $k = 0, 1, 2, \ldots n$. If $a = 2$, $b = 10$, and $n = 4$, find all the values of x_k.

2. Divide the interval $1 \leq x \leq 4$ into six small subintervals. Find the midpoint of each of these subintervals.

3. Suppose $a = 1$ and $b = 4$ and we divide the interval $1 \leq x \leq 4$ into six ($n = 6$) equal length subintervals. Let $f(x) = x^2 + 3x$. Find $f(x_k^*)$ where x_k^* is the midpoint of the kth subinterval. Find an approximate value using $n = 6$ of the area of the region in the xy plane bounded by the lines $x = 1$, $x = 4$, $y = 0$, and the curve $y = x^2 + 3x$. Approximate the value of the area under the curve using long thin rectangles as was done in Example 1.

4. If $f(x) = x^3 + 4x$, $a = 1$, $b = 5$, then consider the expression

$$f\left(a + \frac{2k-1}{2n}(b-a)\right).$$

What is the value of this expression when $k = 7$ and $n = 10$ and when $k = 8$ and $n = 10$.

5. Suppose $f(x) = 4x + 5$, $a = 2$, and $b = 8$, find

a) $a + \dfrac{2k-1}{2n}(b-a)$

b) $f\left(a + \dfrac{2k-1}{2n}(b-a)\right)$

c) $\displaystyle\sum_{k=1}^{n} f\left(a + \frac{2k-1}{2n}(b-a)\right)$

d) $\displaystyle\lim_{n\to\infty} \frac{b-a}{n} \sum_{k=1}^{n} f\left(a + \frac{2k-1}{2n}(b-a)\right).$

This limit gives the area of a certain trapezoid in the plane. Describe this trapezoid.

Hint: $\displaystyle\sum_{k=1}^{n}(2k-1) = n^2.$

6. Find the area of the region in the xy plane bounded by the lines $x = 0$, $x = 5$, the x axis, and the curve $y = x^2$. Hint: If we let $f(x) = x^2$, $a = 0$, and $b = 5$, then the area is given by

$$\lim_{n\to\infty} \frac{b-a}{n} \sum_{k=1}^{n} f\left(a + \frac{2k-1}{2n}(b-a)\right) = \lim_{n\to\infty} \frac{b-a}{n} \sum_{k=1}^{n} f(x_k^*).$$

Hint: $\sum_{k=1}^{n}(2k-1)^2 = \frac{n}{3}(4n^2 - 1).$

6389 The Distance Problem

Suppose we are given an object moving in a straight line with a known variable velocity $v(t)$ with $v(t) > 0$. How do we find the distance which the particle travels during a given time period? If the velocity is constant, then the distance traveled is given by distance = (velocity)(time). If the velocity varies, then it is not so easy to find the distance traveled. If the velocity varies but the time period is very short, we might assume that the velocity does not change very much during this short time period. We could say that the velocity is very nearly constant during this time period. We can then find the distance traveled by the object during a very short time period by multiplying the constant velocity times the time.

The question we wish to consider is: Suppose we have an object moving with a variable velocity, $v(t)$, how far does this object move during the time interval $a \leq t \leq b$? Assume $v(t) > 0$. First, in order to explore the idea let us consider a specific example. Consider the time interval $2 \leq t \leq 10$ with t = time measured in seconds. Divide this time interval into 4 subintervals each 2 seconds long. A good approximation to the value of the velocity during the first small time interval $2 \leq t \leq 4$ should be the value of the velocity at $t = 3$, that is, $v(3)$. Note that $t = 3$ is in the middle of the time interval $2 \leq t \leq 4$. If we multiply the approximate velocity $v(3)$ times the length of time (2 sec), we should get a good approximation to the distance traveled by the object during the first small time interval $2 \leq t \leq 4$.

$$\text{Distance} = (2)v(3).$$

Next approximate the value of velocity during the second small time interval, $4 \leq t \leq 6$ by $v(5)$. This is the velocity at the midpoint time of this time interval. The distance traveled during the second small time interval is approximately $2v(5)$.

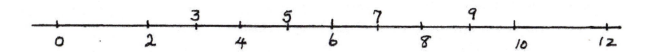

In the same way we can approximate the distance traveled during the third and fourth small time intervals by $(2)v(7)$ and $(2)v(9)$. Thus, the distance traveled by the particle during the time interval $2 \leq t \leq 10$ is approximately the sum of all these distances, that is,

$$2v(3) + 2v(5) + 2v(7) + 2v(9) = 2[v(3) + v(5) + v(7) + v(9)].$$

It is clear that we could get a better approximation to the actual distance traveled by using more time intervals so that each individual interval is shorter. For example, suppose we divide this same time interval $2 \leq t \leq 10$ into 16 small time intervals of equal length. The length of each subinterval of time is $1/2$. The division points on the time line would be

$$t_k = 2 + (k/16)(10 - 2) = 2 + (k/2), \quad k = 0, 1, 2, \ldots 16.$$

The actual values of t_k are

$$t_k = 2, \frac{5}{2}, 3, \frac{7}{2}, 4, \ldots \frac{19}{2}, 10.$$

The first small time interval is $2 \leq t \leq 5/2$. The midpoint of this time interval is $9/4$. The velocity evaluated in the middle of this time interval is $v(9/4)$. The length of the time interval is $1/2$. The approximate distance traveled by the particle during the first short time interval is (time)(velocity) $= (1/2)v(9/4)$. The second small time interval is $5/2 \leq t \leq 3$. The midpoint of this interval is $11/4$. The approximate distance traveled by the particle during the second short time interval is $(1/2)v(11/4)$. Using these time intervals of length $1/2$, we would approximate the total distance traveled by the particle during the time interval $2 \leq t \leq 10$ as the sum of the 16 numbers:

$$\text{distance} = \frac{1}{2}[v(9/4) + v(11/4) + v(13/4) + \ldots + v(37/4) + v(39/4)]$$

This is a better approximation to the distance traveled by the object than the one we got using only 4 short time intervals. The more division points we use to divide the given time interval the better the approximation.

Our idea for computing the distance traveled during a given time interval is to divide the given time interval into short subintervals and approximate the distance traveled during each short subinterval of time. We then add these distances together to get an approximate value for the total distance. The more division points on the time line the better the approximation.

Now that we have this idea for approximating distance traveled let us consider the general case. Suppose an object is moving along a straight line with velocity $v(t)$. If you wish to consider units, you may assume that the units are feet per second. For simplicity let us assume that $v(t) > 0$, that is, the object moves only in the positive direction. Using the method indicated let us approximate the distance traveled by the particle during the time interval $a \leq t \leq b$.

First, we divide the time interval $a \leq t \leq b$ into n short time intervals of equal length. This means that each short time interval is of length

$$\Delta t = \frac{b - a}{n}.$$

The division points on the time line would be

$$t_k = a + k(b - a)/n \text{ where } k = 0, 1, 2, \ldots, n.$$

The first short subinterval of time is all t such that

$$a \leq t \leq a + (b - a)/n.$$

The kth short subinterval of time is all t such that

$$a + (k - 1)\frac{b - a}{n} \leq t \leq a + (k)\frac{b - a}{n} \text{ with } k = 1, 2, \ldots n.$$

The midpoint of the kth time interval is

$$t_k^* = a + \frac{2k - 1}{2n}(b - a) \text{ with } k = 1, 2, 3, \ldots n.$$

We use the velocity at this midpoint time to approximate the velocity during the entire subinterval of time. The approximate distance traveled by the object during the kth short time interval is

$$\text{(time) (velocity)} = \frac{b - a}{n} v\left(a + \frac{2k - 1}{2n}(b - a)\right).$$

275

We add the distance traveled for all n subintervals of time together and get

$$\sum_{k=1}^{n} \frac{b-a}{n} v(a + \frac{2k-1}{2n}(b-a)) = \frac{b-a}{n} \sum_{k=1}^{n} v\left(a + \frac{2k-1}{2n}(b-a)\right).$$

This sum is the approximate distance traveled by the object during the time interval $a \leq t \leq b$. The more frequently we measure the velocity the more accurate the estimates become. This means that the larger the value of n the closer this approximation is to the actual value of distance. This causes us to say that the total distance the object travels during the time interval $a \leq t \leq b$ is obtained by taking larger and larger values of n. The actual distance traveled is the limit as n approaches infinity.

$$\text{distance} = \lim_{n \to \infty} \frac{b-a}{n} \sum_{k=1}^{n} v\left(a + \frac{2k-1}{2n}(b-a)\right).$$

Example 1. The velocity of an object moving in a straight line is given by $v(t) = 3t^2 + 5t$, find the distance traveled by this object during the time interval $0 \leq t \leq 5$.

Solution. We divide the time interval $0 \leq t \leq 5$ into n subdivisions of equal length. The length of each subinterval is $5/n$. The division points on the time line are

$$t_k = a + k\frac{b-a}{n} = 0 + k\frac{5}{n} = \frac{5k}{n}, \ k = 0, 1, 2, \ldots, n.$$

The kth subinterval of time is all values of t such that

$$0 + \frac{k-1}{n}(5) \leq t \leq 0 + \frac{k}{n}(5) \text{ or } \frac{5(k-1)}{n} \leq t \leq \frac{5k}{n}.$$

The midpoint of the kth subinterval of time is

$$t_k^* = \frac{1}{2}\left[\frac{5(k-1)}{n} + \frac{5k}{n}\right] = \frac{5(2k-1)}{2n}.$$

The velocity of the object at the midpoint of the kth subinterval is found by substituting $\frac{5(2k-1)}{2n}$ for t in $v(t) = 3t^2 + 5t$.

$$v\left(\frac{5(2k-1)}{2n}\right) = 3\left[\frac{5(2k-1)}{2n}\right]^2 + 5\left[\frac{5(2k-1)}{2n}\right]$$

$$= \frac{75}{4n^2}(2k-1)^2 + \frac{25}{2n}(2k-1).$$

The approximate distance traveled by the object during the kth subinterval of time is

$$\frac{b-a}{n} v \left(\frac{5(2k-1)}{2n} \right) = \frac{5}{n} \left[\frac{75}{4n^2}(2k-1)^2 + \frac{25}{2n}(2k-1) \right].$$

The sum of the distances over all subintervals of time is

$$\sum_{k=1}^{n} \frac{5}{n} \left[\frac{75}{4n^2}(2k-1)^2 + \frac{25}{2n}(2k-1) \right].$$

$$= \frac{375}{4n^3} \sum_{k=1}^{n}(2k-1)^2 + \frac{125}{2n^2} \sum_{k=1}^{n}(2k-1).$$

Recall that

$$\sum_{k=1}^{n}(2k-1)^2 = \frac{n}{3}(4n^2-1) \text{ and } \sum_{k=1}^{n}(2k-1) = n^2.$$

Substituting we get

$$\frac{b-a}{n} \sum_{k=1}^{n} v \left(a + \frac{2k-1}{2n}(b-a) \right) = \frac{375}{4n^3}\left[\frac{n}{3}(4n^2-1)\right]+\frac{125}{2n^2}[n^2] = \frac{375}{2} - \frac{125}{4n^2}.$$

We find the actual distance traveled by taking the limit as n approaches infinity.

$$\text{distance} = \lim_{n \to \infty} \left[\frac{375}{2} - \frac{125}{4n^2} \right] = \frac{375}{2}.$$

Exercises

1. Suppose $t_k = a+k(b-a)/n$, $k = 0, 1, 2, \ldots n$. If $a = 2$, $b = 5$, and $n = 4$ find the values of the division points t_k. Find the value of t_k^*, the value of t at the midpoint of the kth time interval.

2. Given the time interval $2 \le t \le 12$ divide this interval into $n = 5$ subdivisions using the division points $t_k = a + k(b-a)/n$. Suppose a particle is moving along a line with velocity $v(t) = t^2 + 4t + 3$ ft/sec, use

this subdivision of the time interval $2 \le t \le 12$ to approximate the distance traveled by the particle.

3. Let $v(t) = 2t^2 + 3t$, $a = 0$, and $b = 4$, find the distance traveled by a particle moving with velocity $v(t)$ during the time interval $0 \le t \le 4$ by finding

a) $a + \dfrac{2k-1}{2n}(b-a)$

b) $v\left(a + \dfrac{2k-1}{2n}(b-a)\right)$

c) $\displaystyle\sum_{k=1}^{n} v\left(a + \dfrac{2k-1}{2n}(b-a)\right)$

d) $\dfrac{b-a}{n} \displaystyle\sum_{k=1}^{n} v(a + \dfrac{2k-1}{2n}(b-a))$

e) $\displaystyle\lim_{n \to \infty} \dfrac{b-a}{n} \displaystyle\sum_{k=1}^{n} v(a + \dfrac{2k-1}{2n}(b-a))$.

4. The speedometer reading for a car at the times indicated are given by the following table. Estimate the distance traveled by the car during this 60 seconds of time, $0 \le t \le 60$.

t in seconds	6	18	30	42	54
v in feet/sec	20	24	28	30	33

6391 Definition of the Definite Integral

Suppose we have a region R in the xy plane and the boundaries of this region are the vertical lines $x = a$ and $x = b$, the x axis, and the continuous curve which is the graph of $y = f(x)$. We assume $f(x) > 0$ and $f(x)$ is continuous. We have already found that the area of this region is given by

$$\text{Area} = \lim_{n \to \infty} \frac{b-a}{n} \sum_{k=1}^{n} f\left(a + \frac{2k-1}{2n}(b-a)\right).$$

Suppose we have an object moving on a straight line with variable velocity $v(t)$ with $v(t) > 0$ and $v(t)$ continuous. We have already found that the distance traveled by the object during the time interval $a \le t \le b$ is given by

$$\text{Distance} = \lim_{n \to \infty} \frac{b-a}{n} \sum_{k=1}^{n} v\left(a + \frac{2k-1}{2n}(b-a)\right).$$

Notice that these two limits are exactly the same except that the function is called $f(x)$ in the first limit and the function is called $v(t)$ in the second limit. There are many other problems whose solution involves finding this very same limit. Suppose an object is moved along the x axis from the point where $x = a$ to the point where $x = b$. Suppose a variable force given by $f(x)$ is exerted on the object as it is moved, then the work done by the force is

$$\text{work} = \lim_{n \to \infty} \frac{b-a}{n} \sum_{k=1}^{n} f\left(a + \frac{2k-1}{2n}(b-a)\right).$$

There are a lot of other problems whose solution requires that we find this very same limit. This limit is clearly not easy to evaluate. Therefore, before we look at the problems of finding the area of a region, the distance traveled by an object, or the work done by a force, we first look at just the problem of finding this limit. Once we have mastered finding this limit, then we can consider problems such as finding area where the finding of the limit is just one of the steps. Before we start our detailed study of this limit we give it a name.

Definition of the Definite Integral. If $f(x)$ is a function that is continuous on the interval $a \le x \le b$, then we define the definite integral of $f(x)$

over the interval $a \leq x \leq b$ to be the limit

$$\lim_{n \to \infty} \frac{b-a}{n} \sum_{k=1}^{n} f\left(a + \frac{2k-1}{2n}(b-a)\right).$$

The notation for the definite integral of $f(x)$ over the interval $a \leq x \leq b$ is

$$\int_a^b f(x)dx.$$

This says that

$$\int_a^b f(x)dx = \lim_{n \to \infty} \frac{b-a}{n} \sum_{k=1}^{n} f\left(a + \frac{2k-1}{2n}(b-a)\right).$$

There are several other ways to define the definite integral which are equivalent to this way of defining it. The function $f(x)$ in $\int_a^b f(x)dx$ is called the integrand. The number a is called the lower limit and the number b is called the upper limit of integration. We do not require $f(x) > 0$ in this definition.

Example 1. Find the definite integral

$$\int_1^4 (2x^2 + 4x)dx.$$

Solution. For this example $a = 1$, $b = 4$, and $f(x) = 2x^2 + 4x$. Substituting $a = 1$ and $b = 4$, we get

$$a + \frac{2k-1}{2n}(b-a) = 1 + \frac{3(2k-1)}{2n}.$$

Next we evaluate the function for these values of x.

$$f(1 + 3(2k-1)/2n) = 2[1 + 3(2k-1)/2n]^2 + 4[1 + 3(2k-1)/2n]$$

$$= 2 + 6(2k-1)/n + 18(2k-1)^2/4n^2 + 4 + 6(2k-1)/n$$

$$= 6 + \frac{12(2k-1)}{n} + \frac{9(2k-1)^2}{2n^2}.$$

We need the following sum in order to apply the definition

$$\sum_{k=1}^{n} f\left(a + \frac{2k-1}{2n}(b-a)\right) = \sum_{k=1}^{n}\left[6 + \frac{12(2k-1)}{n} + \frac{9(2k-1)^2}{2n^2}\right]$$

$$= 6\sum_{k=1}^{n}(1) + \frac{12}{n}\sum_{k=1}^{n}(2k-1) + \frac{9}{2n^2}\sum_{k=1}^{n}(2k-1)^2.$$

Clearly $\sum_{k=1}^{n}(1) = n$. Also it is true that

$$\sum_{k=1}^{n}(2k-1) = n^2 \text{ and } \sum_{k=1}^{n}(2k-1)^2 = (n/3)(4n^2-1).$$

Substituting in these values we get that the value of the entire sum is

$$6(n) + (12/n)(n^2) + (9/2n^2)(n/3)(4n^2-1) = 18n + 6n - (3/2n)$$

The definite integral is given by the following limit:

$$\int_{1}^{4}(2x^2 + 4x)dx = \lim_{n\to\infty}\frac{3}{n}[24n - \frac{3}{2n}] = \lim_{n\to\infty}72 - \frac{9}{2n^2} = 72.$$

Example 2. Evaluate the definite integral $\int_{0}^{5} x^3 dx$.

Solution. Because we have no theorems to help us we must apply the definition of the definite integral in order to evaluate this definite integral. For this example $a = 0$, $b = 5$, and $f(x) = x^3$.

$$a + \frac{2k-1}{2n}(b-a) = 0 + \frac{5(2k-1)}{2n}.$$

$$f\left(\frac{5(2k-1)}{2n}\right) = \left[\frac{5(2k-1)}{2n}\right]^3 = \frac{125(2k-1)^3}{8n^3}$$

$$\frac{b-a}{n}\sum_{k=1}^{n} f\left(a + \frac{2k-1}{2n}(b-a)\right) = \frac{5}{n}\sum_{k=1}^{n}\frac{125(2k-1)^3}{8n^3} = \frac{625}{8n^4}\sum_{k=1}^{n}(2k-1)^3.$$

It is true that $\sum_{k=1}^{n}(2k-1)^3 = n^2(2n^2-1)$.

Substituting for the value of the sum, we have

$$\int_0^5 x^3\,dx = \lim_{n\to\infty} \frac{625}{8n^4}[n^2(2n^2-1)] = \lim_{n\to\infty} \frac{625}{8}(2-\frac{1}{n^2}) = \frac{625}{4}.$$

Exercises

1. Given the integral $\int_1^6 (6x^2 + 4x)dx$, find

a) $a + \dfrac{2k-1}{2n}(b-a)$

b) $f(a + \dfrac{2k-1}{2n}(b-a))$

c) $\sum_{k=1}^{n} f(a + \dfrac{2k-1}{2n}(b-a))$

d) Simplify $\dfrac{b-a}{n} \sum_{k=1}^{n} f(a + \dfrac{2k-1}{2n}(b-a))$ and find the limit

$\lim_{n\to\infty} \dfrac{b-a}{n} \sum_{k=1}^{n} f(a + \dfrac{2k-1}{2n}(b-a))$.

Calculate the following definite integrals using the definition.

2. $\int_0^8 x^3\,dx$

3. $\int_1^6 (x^2 + 5x)dx$

4. $\int_1^5 (2x^2 - 6x)dx$

6395 The Fundamental Theorem of Calculus

Example 1. Evaluate the definite integral $\int_2^{10} x^3 dx$.

Solution. The definition of definite integral tells us that

$$\int_2^{10} x^3 dx = \lim_{n \to \infty} \frac{8}{n} \sum_{k=1}^{n} [2 + 4(2k-1)/n]^3.$$

This is not an easy limit to find. However, it turns out that there is an easier way to find the value of definite integrals that does not involve evaluating these hard limits. We have a theorem which makes finding the value of a definite integral much easier. We do not have to evaluate this awful limit.

Fundamental Theorem of Calculus. If $f(x)$ is continuous for $a \le x \le b$ and $F'(x) = f(x)$ on the interval $a \le x \le b$, then

$$\int_a^b f(x) dx = F(b) - F(a).$$

This theorem is fairly difficult to prove and so we will not give the proof here. Note that we are assuming that $f(x)$ is defined and continuous for $a \le x \le b$ and that $F'(x) = f(x)$ for $a \le x \le b$.

Let us use this theorem to evaluate the definite integral $\int_2^{10} x^3 dx$. In order to do this, we need a function $F(x)$ such that $F'(x) = x^3$. This says that $F(x)$ must be an antiderivative of x^3. The easiest antiderivative of x^3 is $x^4/4$. Any antiderivative will do. We can now apply the Fundamental Theorem of Calculus to evaluate the definite integral.

$$\int_2^{10} x^3 dx = \frac{x^4}{4} \Big|_2^{10} = \frac{(10)^4}{4} - \frac{2^4}{4} = 2496.$$

This means that

$$\lim_{n \to \infty} \frac{8}{n} \sum_{k=1}^{n} [2 + 4(2k-1)/n]^3 = 2496.$$

We use the notation $F(x) \Big|_a^b$ to denote $F(b) - F(a)$.

283

The Fundamental Theorem of Calculus requires that $F'(x) = f(x)$ for $a \leq x \leq b$. This implies that $F'(x)$ exists for $a \leq x \leq b$. This may mean that $F'(a)$ and $F'(b)$ are one sided derivatives. The definition of the definite integral does not require that $f(x)$ be continuous for $a \leq x \leq b$. In order to apply the Fundamental Theorem of Calculus it is required that $f(x)$ be continuous for $a \leq x \leq b$. This means that there are functions $f(x)$ such that $\int_a^b f(x)dx$ exists, but we are unable to find the value of $\int_a^b f(x)dx$ using the Fundamental Theorem of Calculus. The Fundamental Theorem of Calculus and the Definition of Integral both require that the numbers a and b are finite. The definite integral $\int_a^b f(x)dx$ is called a proper integral if it can be evaluated using the Fundamental Theorem of Calculus. This means $F'(x) = f(x)$ for $a \leq x \leq b$ and $f(x)$ is continuous for $a \leq x \leq b$.

Example 2. Evaluate the definite integral $\int_1^4 (5x^4 + 7x^2)dx$.

Solution. We want to use the Fundamental Theorem of Calculus. In order to do this we need an antiderivative of $5x^4 + 7x^2$. The simplest antiderivative of $5x^4 + 7x^2$ is $x^5 + (7/3)x^3$. Knowing this antiderivative the Fundamental Theorem of Calculus tells us that

$$\int_1^4 (5x^4 + 7x^2)dx = x^5 + (7/3)x^3 \Big|_1^4 = (4)^5 + (7/3)(4)^3 - [1 + 7/3] = 1170.$$

Example 3. Evaluate the definite integral $\int_{-2}^3 x^{-2}dx$.

Solution. We want to use the Fundamental Theorem of Calculus. An antiderivative of $f(x) = x^{-2}$ is $F(x) = x^{-1}$. However, $F(x) = x^{-1}$ is not defined for $x = 0$. This says that $F'(x) = f(x)$ is not true for $x = 0$. Since $-2 < 0 < 3$ the Fundamental Theorem of Calculus does not apply to this problem. This is not a proper integral. In fact, let us look back at the Fundamental Theorem of Calculus. It says that in order to evaluate the integral $\int_{-2}^3 f(x)dx$ using the Fundamental Theorem of Calculus it must be true that $f(x)$ is a continuous function on the interval $-2 \leq x \leq 3$. Since $f(x) = x^{-2}$ is not continuous at $x = 0$, the definite integral

$$\int_{-2}^3 x^{-2}dx$$

is not a proper integral and is not defined since our definition of definite integral requires that the integrand of a definite integral be continuous.

Definition. If $a < b$, then we define

$$\int_b^a f(x)dx = -\int_a^b f(x)dx.$$

Interchanging the limits of integration reverses the sign on the integral. The following theorems give some useful properties of the definite integral. These theorems are all proved by working with the definition of the definite integral.

Theorem 1. If $f(x) > 0$ for $a \leq x \leq b$, then $\int_a^b f(x)dx > 0$.

Theorem 2. $\int_a^b [cf(x)]dx = c \int_a^b f(x)dx$.

Theorem 3. $\int_a^b [f(x) + g(x)]dx = \int_a^b f(x)dx + \int_a^b g(x)dx$

Theorem 4. If $a < c < b$, then $\int_a^c f(x)dx + \int_c^b f(x)dx = \int_a^b f(x)dx$.

Remark. In proving the Fundamental Theorem of Calculus we assume that $f(x)$ is defined and continuous for $a \leq x \leq b$ and that $F'(x) = f(x)$ for $a \leq x \leq b$. This means that we can not find the value of the integral

$$\int_0^1 \frac{dx}{\sqrt{1 - x^2}}$$

using the Fundamental Theorem of Calculus. The reason is that the integrand $\dfrac{1}{\sqrt{1 - x^2}}$ is not continuous at $x = 1$.

Exercises

Use the Fundamental Theorem of Calculus to evaluate the following definite integrals.

1. $\displaystyle\int_1^4 (8x^3 + 12x^2)dx$

2. $\displaystyle\int_0^1 \frac{dx}{1 + x^2}$

3. $\displaystyle\int_0^{\sqrt{3}/2} \frac{dx}{\sqrt{1 - x^2}}$

285

4. $\displaystyle\int_0^{\pi/4} \cos 2x\,dx$ 5. $\displaystyle\int_1^5 \frac{dx}{x^2}$ 6. $\displaystyle\int_0^{\sqrt{21}} x\sqrt{x^2+4}\,dx$

7. $\displaystyle\int_0^7 (x+3)^4\,dx$ 8. $\displaystyle\int_0^2 \frac{x\,dx}{\sqrt{3x^2+4}}$

9. Explain why the integral $\displaystyle\int_0^1 x^{-1/2}$ can not be evaluated using the Fundamental Theorem of Calculus. Explain why the definite integral $\int_0^2 \frac{dx}{\sqrt{4-x^2}}$ can not be evaluated using the Fundamental Theorem of Calculus as $\arcsin(x/2)\,|_0^2$.

6397 Area and Velocity

Recall from our earlier work that the area of the region R which is bounded by the curve $y = f(x)$, the vertical lines $x = a$ and $x = b$, and the x axis is given by

$$\text{Area} = \lim_{n \to \infty} \frac{b-a}{n} \sum_{k=1}^{n} f\left(a + \frac{2k-1}{2n}(b-a)\right).$$

Also we define the definite integral to be this very same limit. It follows that the following theorem is true.

Theorem 1. Suppose $f(x)$ is a continuous function and $f(x) > 0$ for $a \le x \le b$. Given the region R which is bounded by the curve which is the graph of $y = f(x)$, the vertical lines $x = a$ and $x = b$, and the x axis, then the area of the region R is given by the definite integral

$$\int_{a}^{b} f(x)dx.$$

Example 1. Find the area of the region R which is bounded by the graph of $y = 16 - x^2$, the vertical lines $x = -2$ and $x = 3$, and the x axis.

Solution. Although it is not absolutely necessary for this example, it is almost always a good idea to make a sketch of the region when asked to find its area.

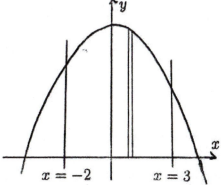

For this example $a = -2$, $b = 3$, and $f(x) = 16 - x^2$. Theorem 1 then says that the area is equal to the value of the definite integral

$$\int_{-2}^{3} (16 - x^2)dx.$$

An antiderivative of $16 - x^2$ is $16x - (1/3)x^3$. The Fundamental Theorem of Calculus says that the value of this integral is

$$16x - (1/3)x^3 \Big|_{-2}^{3} = [16(3) - (1/3)(3)^3] - [16(-2) - (1/3)(-2)^3] = 205/3.$$

The area under the curve is $\frac{205}{3}$ square units. These units are commonly not given a name such as square feet or square meters. We assume that the x and y axis are both marked off in the same units, then the units of area are these units squared. If the x and y axis are marked in feet or meters, then the units of area would be square feet or square meters. We do not discuss area when the x axis is marked in one unit say feet and the y axis is marked in another unit, say meters. This means that calculators do not usually show area.

Suppose we have an object moving along the x axis with velocity $v(t)$ where $v(t) > 0$. In our previous work we found that the distance traveled by the object during the time interval $a \leq t \leq b$ is given by

$$\text{distance} = \lim_{n \to \infty} \frac{b-a}{n} \sum_{k=1}^{n} v\left(a + \frac{2k-1}{2n}(b-a)\right).$$

We defined the definite integral $\int_a^b v(t)dt$ to be this very same limit. It follows that the following theorem is true.

Theorem 2. Suppose that $v(t) > 0$. If an object is moving along the x axis with velocity $v(t)$, then the distance traveled by the object during the time interval $a \leq t \leq b$ is equal to the value of the definite integral

$$\int_a^b v(t)dt.$$

Example 2. A particle is moving along the x axis with velocity $v(t) = 3t^2 + 8t + 11$. Find the distance traveled by the object during the time interval $2 \leq t \leq 10$.

Solution. A direct application of Theorem 2 says that the distance is given by

$$\int_2^{10} (3t^2 + 8t + 11)dt.$$

An antiderivative of $3t^2+8t+11$ is t^3+4t^2+11t. The Fundamental Theorem of Calculus tells us that the value of this integral is

$$t^3 + 4t^2 + 11t\Big|_2^{10} = [10^3 + 4(10^2) + 11(10)] - [2^3 + 4(2^2) + 11(2)] = 1464.$$

The object moved 1464 units. When an object is moving along the x axis we assume that the units for measuring distance are the units on the x axis. These units are commonly not given a name such as feet or meters. We say that an object moves 1464 meaning that it moves 1464 of these units. If an object is moving along a straight line marked in units of feet or meters, then the answer should be given with units of feet or meters. We would say that the object moves 1464 feet or 1464 meters.

Exercises

1. Find the area of the region bounded by the curve which is the graph of the equation $y = 8x - x^2$, the vertical lines $x = 2$ and $x = 8$, and the x axis.

2. Find the area of the region bounded above by the curve whose equation is $y = (-2 + x)(10 - x)$. The region is also bounded by the vertical lines $x = 4$ and $x = 10$ and the x axis.

3. Find the area of the region bounded above by the curve $y = \dfrac{4}{1 + x^2}$, the vertical lines $x = 4$ and $x = 10$, and the x axis.

4. An object is moving along the x axis with velocity $v(t) = 4t^2 - 20t + 29$. Find the distance traveled by the object during the time interval $1 \le t \le 5$.

5. A car starts moving along a long straight road at a certain point. Distance along the road is measured in meters. The velocity of the car is given by $v(t) = (1/2)t + 5$ meters per second. Find the distance traveled by the car during the time interval $0 \le t \le 10$. Note that time is measured in seconds.

Example 1. Let R denote the region bounded above by the parabola whose equation is $y = 16 - x^2$, below by the line $y = -2x + 8$. Find the area of this region.

Solution. Theorem 1 in the previous section (6397) is the only theorem we have had up to this point that enables us to find the area of a region in the xy plane. Theorem 1 requires that $f(x) > 0$ and that the lower boundary of the region must be the x axis. We can not solve this problem directly using Theorem 1. The reason we can not use Theorem 1 directly to find the area of this region is that this region is not bounded below by the x axis. However, we can still use Theorem 1 to solve this problem by considering two regions each of which is bounded below by the x axis.

First, we need to be considering a region which is between two vertical lines. What are the two vertical lines for this problem? In order to find where the line and parabola intersect, we solve the equations $y = -2x + 8$ and $y = 16 - x^2$ simultaneously.

$$-2x + 8 = 16 - x^2$$
$$x^2 - 2x - 8 = 0$$
$$x = 4 \text{ or } x = -2.$$

We conclude that the given line and parabola intersect at the point where $x = 4$ and where $x = -2$. This means that the vertical lines $x = -2$ and $x = 4$ pass through the point where the parabola $y = 16 - x^2$ and the line $y = -2x + 8$ intersect. Let us make a sketch.

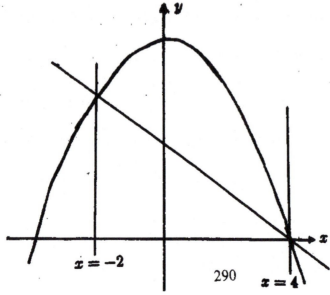

$x = -2$

290

$x = 4$

The vertical line $x = -2$ can be taken as the vertical line which forms the boundary of the region R on the left. The vertical line $x = 4$ can be taken as the vertical line which forms the boundary of the region R on the right.

Let R_1 denote the region bounded by the parabola whose equation is $y = 16 - x^2$, the vertical line $x = -2$, the vertical line $x = 4$, and the x axis. Let R_2 denote the region bounded by the line $y = -2x + 8$, the vertical lines $x = -2$ and $x = 4$, and the x axis.

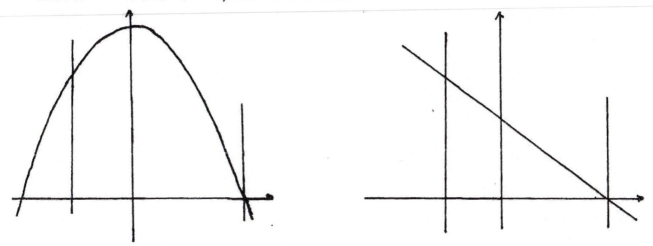

It is clear that the given region R is the region R_1 take away the region R_2. Therefore

$$\text{Area of } R = (\text{Area of } R_1) - (\text{Area of } R_2)$$

Note that both the regions R_1 and R_2 are bounded below by the x axis. We can find the areas of both the regions R_1 and R_2 as a direct application of Theorem 1. Applying Theorem 1, we get

$$\text{Area of } R_1 = \int_{-2}^{4} (16 - x^2)\,dx$$

$$\text{Area of } R_2 = \int_{-2}^{4} (-2x + 8)\,dx.$$

It follows that

$$\text{Area of } R = \int_{-2}^{4} (16 - x^2)\,dx - \int_{-2}^{4} (-2x + 8)\,dx$$

$$= \int_{-2}^{4} [(16 - x^2) - (-2x + 8)]\,dx.$$

Theorem 1. Suppose $f(x) > g(x) > 0$ for $a \le x \le b$. Let R denote the region which is bounded above by the curve which is the graph of $y = f(x)$, below by the curve which is the graph of $y = g(x)$, and is between the vertical lines $x = a$ and $x = b$, $a < b$, then the area of R is given by the definite integral

$$\int_a^b [f(x) - g(x)]dx.$$

We prove this theorem using the idea that was used to work the previous example.

Example 2. Find the area of the region bounded by the curve $y = x^2 - 9$, the vertical lines $x = -2$ and $x = 3$, and the x axis.

Solution. The curve $y = x^2 - 9$ crosses the x axis at the points where $x = -3$ and where $x = 3$. Let us call the given region R_1. The region R_1 is bounded on the right by the line $x = 3$ and on the left by $x = -2$. Let us make a sketch of the region.

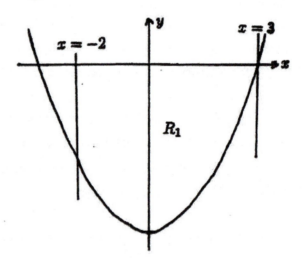

We note that the function $f(x) = x^2 - 9$ does not satisfy the condition $f(x) > 0$ for $-2 \le x \le 3$. Since the condition $f(x) > 0$ is not satisfied we can not directly apply Theorem 1 to find the area of this region. But we do have a trick for getting around this. Let us consider the curve which is the graph of the function $g(x) = -f(x) = 9 - x^2$. The curve $g(x) = -(x^2 - 9)$ is a reflection of the given curve $f(x) = x^2 - 9$ through the x axis. Let R_2 denote the region bounded by the curve $g(x) = 9 - x^2$, the vertical lines $x = -2$ and $x = 3$, and the x axis.

292

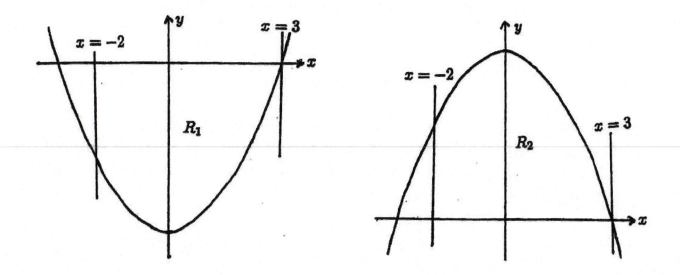

The region R_2 is an exact reflection of the region R_1 through the x axis. The area of the new region R_2 is exactly the same as the area of the given region R_1. We can compute the area of R_1 by computing the area of R_2. For the region R_2 we have $g(x) > 0$. Theorem 1 enables us to say that the area of R_2 is

$$\int_{-2}^{3} g(x)\,dx = -\int_{-2}^{3} (x^2 - 9)\,dx.$$

The area of the given region R_1 is

$$\text{Area} = (-1)\int_{-2}^{3} (x^2 - 9)\,dx.$$

Using a discussion similiar to the one used to solve this problem we conclude that the following theorem is true.

Theorem 2. Suppose a region R is bounded by the curve $y = f(x)$ with $f(x) < 0$ for $a \le x \le b$, the vertical lines $x = a$ and $x = b$, and the x axis, then the area of R is given by the definite integral

$$(-1)\int_{a}^{b} f(x)\,dx.$$

Example 3. Let R denote the region bounded above by the line $y = -2x$ and below by the parabola $y = x^2 - 10x$. Find the area of R.

Solution. First, we find the point where the line $y = -2x$ and the parabola $y = x^2 - 10x$ intersect by solving the equations simultaneously.

$$x^2 - 10x = -2x$$
$$x = 0 \text{ and } x = 8.$$

The vertical line $x = 0$ (the y axis) passes through the point $(0, 0)$ where the line and parabola intersect. The vertical line $x = 8$ passes through the point $(8, -16)$ where the line and parabola intersect. As usual we sketch a graph

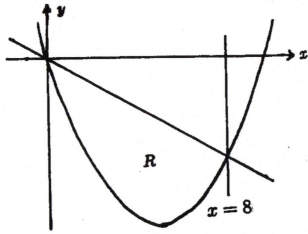

The region R is bounded on the left by the line $x = 0$ and on the right by the line $x = 8$. Let R_1 denote the region bounded by the x axis, the lines $x = 0$ and $x = 8$, and the parabola $y = 10x - x^2$. Let R_2 denote the region bounded by the x axis, the lines $x = 0$ and $x = 8$, and the line $y = -2x$.

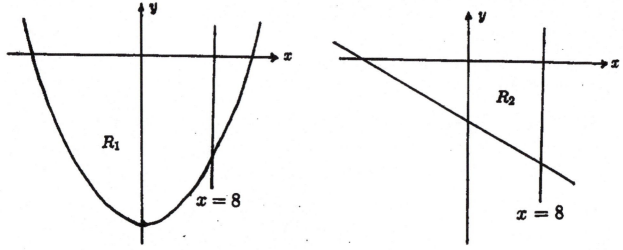

It is clear that the region R is the region R_1 take away the region R_2.

$$\text{Area of } R = (\text{Area of } R_1) - (\text{Area of } R_2).$$

Note that all these regions are below the x axis. We can make a direct

294

application of Theorem 2 to get

$$\text{Area of } R_1 = (-1)\int_0^8 (x^2 - 10x)dx$$

$$\text{Area of } R_2 = (-1)\int_0^8 (-2x)dx$$

$$\text{Area of } R = (-1)\int_0^8 (x^2 - 10x)dx - (-1)\int_0^8 (-2x)dx$$

$$= \int_0^8 [(-2x) - (x^2 - 10x)]dx.$$

Using a discussion similar to the one used to solve this problem, we conclude that the following theorem is true.

Theorem 3. Suppose $f(x) > g(x)$, but $f(x) < 0$ and $g(x) < 0$ for $a \leq x \leq b$. Let R denote the region which is bounded above by the curve which is the graph of $y = f(x)$, below by the curve which is the graph of $y = g(x)$, and is between the vertical lines $x = a$ and $x = b$, $a < b$, then the area of R is given by

$$\int_a^b [f(x) - g(x)]dx.$$

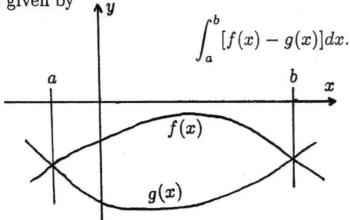

We are now able to state a general rule which is good for finding areas that are both above and below the x axis. In order to prove that this rule is true in all circumstances we need to consider a few more cases. However, based on what we have already done it is easy to show that the theorem is correct in these cases.

Theorem 4. Let R denote the region which is bounded above by the curve which is the graph of $y = f(x)$ and below by the curve which is the graph of $y = g(x)$ and which is between the vertical lines $x = a$ and $x = b$, $a < b$, then the area of the region R is given by the definite integral

$$\int_a^b [f(x) - g(x)]dx.$$

Example 4. Find the area of the region bounded by the parabola $y = 16 - x^2$ and the line $y = -2x + 1$.

Solution. Where are the vertical lines? We need to find the points where the parabola $y = 16 - x^2$ and the line $y = -2x + 1$ intersect. We solve the equations $y = 16 - x^2$ and $y = -2x + 1$ simultaneously in order to find these points.

$$-2x + 1 = 16 - x^2$$
$$x^2 - 2x - 15 = 0$$
$$(x - 5)(x + 3) = 0$$
$$x = 5 \text{ or } x = -3.$$

The curves intersect at the point where $x = -3$, that is, $(-3, 7)$, and at the point where $x = 5$, that is, $(5, -9)$. The region in question is between the vertical lines $x = -3$ and $x = 5$. A sketch of the region is helpful.

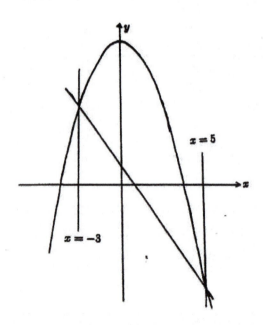

It is now easy to apply the Theorem 4 to find the area. The equations of the vertical lines always give the limits on the integral and the integrand is always top curve minus bottom curve.

$$\text{Area} = \int_{-3}^{5} [(16 - x^2) - (-2x + 1)]dx$$

$$= \int_{-3}^{5} [-x^2 + 2x + 15]dx.$$

An antiderivative of $-x^2 + 2x + 15$ is $(-1/3)x^3 + x^2 + 15x$. The value of the integral is

$$-\frac{x^3}{3} + x^2 + 15x \Big|_{-3}^{5} = -\frac{125}{3} + 25 + 75 - \left[\frac{27}{3} + 9 - 45\right] = \frac{256}{3}.$$

We have a method for finding the area of a region R which is bounded above and below by curves and is between two vertical lines. We can extend our ideas for finding area to enable us to find the area of any region bounded by smooth curves. For example, we can easily find the area of a region bounded on the left and right by smooth curves and on the top and bottom by horizontal lines. In order to find the area of such a region we need only choose y as the independent variable instead of x. The easiest way to handle this is to rewrite the functions involved using x as the independent variable. This would enable us to follow the usual practice of graphing the independent variable along the horizontal axis.

Example 5. Find the area of the region bounded by the curve which is the graph of $x = y^2 - 2y - 3$ and the line $x = 5$.

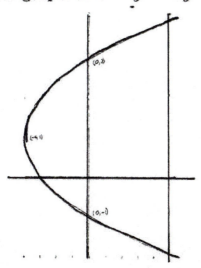
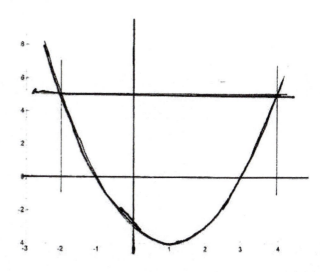

Solution. First, we can solve this problem as we have always solved find the area problems by using x as the independent variable. The graph of $x = y^2 - 2y - 3$ is a parabola, the top half of the parabola is the graph of $y = 1 + \sqrt{x+4}$ for $x \geq -4$ and the bottom half of the parabola is the graph of $y = 1 - \sqrt{x-4}$ for $x \geq -4$. The area of the region is given by

$$\int_{-4}^{5} \left[1 + \sqrt{x+4}\right] - \left[1 - \sqrt{x+4}\right] \, dx = 2 \int_{-4}^{5} \sqrt{x+4} \, dx$$

$$2 \left[\frac{2}{3}(x+4)^{3/2} \,|_{-4}^{5}\right] = \frac{4}{3}(9)^{3/2} = 36.$$

Second, we can more easily solve this problem using familiar ideas. We just restate the problem using x as the independent variable. The problem restated: find the area of the region bounded by the parabola $y = x^2 - 2x - 3$ and the line $y = 5$. The graph of $y = x^2 - 2x - 3$ and $y = 5$ cross at $x = -2$ and $x = 4$. The area is

$$\int_{-2}^{4} \left[5 - (x^2 - 2x - 3)\right] \, dx = -\frac{x^3}{3} + x^2 + 8x \,|_{-2}^{4} = 36$$

Exercises

1. Find the area of the region bounded by the curve $y = x^2 - 6x$ and the x axis.

2. Find the area of the region bounded by the curve $y = x^2 - 8x$, the vertical lines $x = 2$ and $x = 10$, and the x axis.

3. Find the area of the region bounded by the curve which is the graph of $y = 20 - x^2$ and the curve which is the graph of $y = x^2 + 2$.

4. Find the definite integral which when evaluated will equal to the area of the region bounded by the parabola $y = x^2 - 4$ and the line $y = 2x - 1$.

5. Find the definite integral when evaluated will equal to the area of the region bounded by the parabola $y = 2x - x^2$ and the parabola $y = x^2 - 4x$.

6. Find the area of the region bounded by the curve which is the graph of $y = x^2$ and the graph of the curve of $y = \frac{2}{x^2+1}$.

298

7. Find the area of the region bounded by the curve $y = \sin x$ and the curve $y = \cos x$ and which is between the vertical lines $x = 0$ and $x = \pi$.

8. Find the area of the two part finite region bounded by the parabola $y = x^2 - 3x$, the line $y = x$, and the line $x = 6$.

9. Find the area of the region which is bounded by the graph of $x = -y^2 + 4y$ and the graph of $x = y^2 - 2y - 8$.

6402 Velocity

Example 1. Suppose an object is moving along a straight line with distance measured in feet. Find the distance that the object moves during the time interval $2 \leq t \leq 5$ when the velocity is given by $v(t) = t^2 - 6t + 10$ ft/min.

Solution. The statement of the problem does not give us the position of the object at time $t = 0$ nor does the statement enable us to determine the position of the object at any other time. In our previous work we concluded that when $v(t) > 0$ the distance traveled by an object during the time interval $a \leq t \leq b$ is given by $\int_a^b v(t)dt$. Note that $t^2 - 6t + 10 > 0$ for all values of t. The distance traveled by the object is

$$\int_2^5 (t^2 - 6t + 10)dt = \frac{t^3}{3} - 3t^2 + 10t \Big|_2^5 = 6.$$

We conclude that the object was displaced 6 feet in the positive direction. We are moving along a straight line which has a positive direction. Clearly it is possible that an object can move in the negative direction as well as in the positive direction.

Example 2. Suppose an object is moving along the x axis with variable velocity $v(t) = 3t^2 - 12t$. Find the distance traveled by this object during the time interval $0 \leq t \leq 4$.

Solution. In our previous problems about distance traveled we always assumed that $v(t) > 0$. Suppose $0 \leq t \leq 4$, then $t-4 \leq 0$ and $t > 0$. It follows that $t(t - 4) \leq 0$ or $3t(t - 4) \leq 0$. This says that $v(t) \leq 0$ for $0 \leq t \leq 4$. The object in this example is moving in the negative direction. But, we can still ask the question: how far does the object move? Distance moved is a positive number. Since time change is a positive number we need a positive number for velocity in order to get a positive number for distance. In order to compute distance traveled we consider only the magnitude of the velocity. The fact that a velocity is negative just indicates that it will result in a movement in the negative direction. In order to compute distance traveled we use $|3t^2 - 12t| = (-1)(3t^2 - 12t)$ as the velocity

$$\textbf{distance traveled} = (-1) \int_0^4 (3t^2 - 12t)dt = 32.$$

There are really two different questions we ask about the movement of an object along a straight line. In order to make the meaning of our question clear we state two definitions.

Definition of displacement. (Net change of position) Suppose an object is in the position $x = x_1$ when $t = a$ and in the position $x = x_2$ when $t = b$, then the displacement of the object during the time interval $a \leq t \leq b$ is $x_2 - x_1$. Note that displacement can be positive or negative.

Suppose an object is moving along a line with velocity $v(t)$, then displacement during the time interval $a \leq t \leq b$ is given by $\int_a^b v(t)dt$.

Definition of distance traveled. The distance moved by an object moving on a straight line is the total distance. This means that a movement in the negative direction also results in a positive distance traveled.

This means that a negative velocity must be counted as positive when we compute distance traveled. The negative in a negative velocity $v(t)$ indicates the object is moving in the negative direction, but to compute distance traveled which is positive we must place a negative sign in front of the given $v(t)$. In order to compute distance we must use $|v(t)| = [-v(t)]$ as the velocity when $v(t) < 0$ in order to get a positive value for distance.

Example 3. A particle is moving along the x axis with velocity $v(t) = t^2 - 8t + 12$. Find both the displacement and total distance traveled by this particle during the time interval $0 \leq t \leq 10$.

Solution. The displacement is easy. We just find the amount of change of position. This is given by

$$\int_0^{10} (t^2 - 8t + 12)dt = \frac{160}{3}.$$

The object changed its position by 160/3 units in the positive direction. In order to find the distance traveled we need to find the values of t for which $v(t)$ is positive and the values of t for which $v(t)$ is negative. Let us sketch a graph of $v(t) = (t - 2)(t - 6)$.

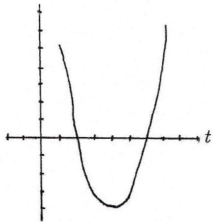

From the graph we see that $v(t) \geq 0$ for $0 \leq t \leq 2$ and for $6 \leq t \leq 10$, but $v(t) \leq 0$ for $2 \leq t \leq 6$. The distance traveled by the object during the time interval $2 \leq t \leq 6$ is

$$(-1)\int_2^6 (t^2 - 8t + 12)dt.$$

Thus, the distance traveled by the particle during the total time interval $0 \leq t \leq 10$ is given by the sum of the three integrals

$$\int_0^2 (t^2 - 8t + 12)dt + (-1)\int_2^6 (t^2 - 8t + 12)dt + \int_6^{10} (t^2 - 8t + 12)dt.$$

$$= \frac{32}{3} + \frac{32}{3} + \frac{160}{3} = \frac{224}{3}.$$

We can also express the distance traveled by this particle as

$$\int_0^{10} |t^2 - 8t + 12|dt.$$

The absolute value means we are always taking the positive value of velocity.

In general if an object is moving with velocity $v(t)$ during the time interval $a \leq t \leq b$, then

$$\text{displacement} = \int_a^b v(t)dt$$

$$\text{distance traveled} = \int_a^b |v(t)|dt.$$

This seems a good time to talk about the integral $\int_a^b |f(x)|dx$ in general. We do not have any antiderivative formulas that involve the absolute value sign. In order to evaluate the integral $\int_a^b |f(x)|dx$ we must get rid of the absolute value signs.

Example 4. Find $\int_0^{2\pi} |\sin x|dx$.

Solution. In order to get rid of the absolute value signs we need to find the values of x such that $\sin x \geq 0$ and the values of x such that $\sin x \leq 0$ for $0 \leq x \leq 2\pi$. We know

$$\sin x \geq 0 \quad \text{for } 0 \leq x \leq \pi$$
$$\sin x \leq 0 \quad \text{for } \pi \leq x \leq 2\pi.$$

This means that

$$|\sin x| = \sin x \text{ for } 0 \leq x \leq \pi$$
$$|\sin x| = (-1)\sin x \text{ for } \pi \leq x \leq 2\pi.$$

This means that

$$\int_0^{2\pi} |\sin x|dx = \int_0^{\pi} (\sin x)dx + \int_{\pi}^{2\pi} (-1)(\sin x)dx$$
$$= -\cos x \Big|_0^{\pi} + \cos x \Big|_{\pi}^{2\pi}$$
$$= -[-1-1] + [1-(-1)] = 4.$$

Exercises

1. Evaluate the integrals $\int_0^8 (5t - t^2)dt$ and $\int_0^8 |5t - t^2|dt$.

2. Suppose an object is moving on a straight line with variable velocity $v(t) = t^2 - 6t$. Find both the displacement and total distance traveled for this object during the time interval $0 \leq t \leq 10$.

3. A particle is moving on the x axis with velocity given by $v(t) = -3t^2 + 18t - 15$. Find both the displacement (net change of position) and total distance traveled for this particle during the time interval $0 \leq t \leq 8$.

4. A car is traveling along a straight road at 60 mph. Starting at a certain time it decelerates at a constant rate of 10 miles/hr^2. How long before the

car's speed is 45 mph? How far does the car travel before it comes to a stop?

5. Evaluate the integral $\int_{-2}^{8} |6x - x^2| dx$.

6403 Volume by Disks

In our future study will be able to find the volume of any solid using the calculus of two independent variables, but for now we are studying the calculus of one independent variable. Even so we are still able to find the volume of certain solids using the calculus of one independent variable. These solids have a special shape and are called solids of revolution. We begin our study by describing what we mean by a solid of revolution. In order to generate a solid of revolution we start with a figure in the plane. We start with a closed bounded region in the plane such as a circle or a rectangle. We also have a line in this plane which does not pass through the interior of the given region. We then form a three dimensional figure (a solid) by revolving the region about the line. As the region revolves about the line it traces out a solid. Let us consider the case when the region we start with is a rectangle and the line is one side of the rectangle.

Revolve the rectangle about the line using the line as an axis. This generates a solid which is a circular cylinder.

Let us consider the case when the region we start with is a semicircle and the line is the flat side of the semicircle extended.

Revolving the semicircle about this line generates a solid which is a sphere.

Let us consider the case when the region we start with is a circle and the line is a line which does not touch the circle

Revolving the circle about the line generates the solid which is a doughnut (torus).

We will use the xy plane as a plane in which to draw regions and

lines. We use the xy plane since we need equations in order to describe the boundaries of the regions. The simplest problems are those for which the x axis is the line about which we revolve the region.

Suppose the region R is a long thin rectangle with its short side part of the x axis.

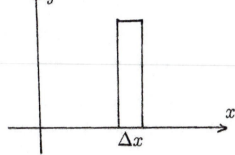

Suppose the width of the rectangle is Δx and the height is y. What is the volume of the solid generated when this thin rectangle R is revolved about the x axis. Let S denote this solid. This solid is a very short cylinder. The radius of the base of the cylinder is y and the height of the cylinder is Δx. The volume of the cylinder is

$$\text{Volume} = \pi(\text{radius})^2(\text{height}) = \pi y^2 \Delta x.$$

We will need this formula because we will need to find the volume of solids composed of several short cylinders.

Example 1 Let R denote the region which is bounded by the curve which is the graph of $y = 6x - x^2$ and the x-axis. Find the volume of the solid generated when the region R is revolved about the x axis.

Solution. The curve $y = 6x - x^2$ crosses the x axis at the point where $x = 0$ and the point where $x = 6$. The curve is a parabola. This enables us to sketch the region R.

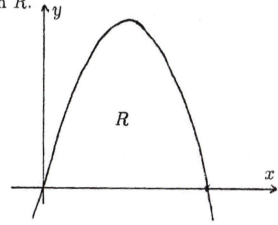

Let S denote the solid generated when the region R is revolved about the x-axis. The solid S has the x axis as an axis of symmetry. The solid S looks something like a flattened sphere. We are going to find the volume of this solid by slicing it up like a loaf of bread and finding the volume of each slice. Slices are made by cutting the solid perpendicular to the x axis. We then add together the volumes of all the slices to get the volume of the whole loaf.

We are going to approximate the volume of each slice of the solid S with a short cylinder. We begin by forming long thin rectangles in the region R in the same way that we did when finding area. Since the curve $y = 6x - x^2$ crosses the x axis at $x = 0$ and $x = 6$, we consider the interval $0 \le x \le 6$. Divide the interval $0 \le x \le 6$ into small subintervals of length $\Delta x = 6/n$. Let x_{k-1} and x_k denote the end points for the kth subinterval, that is, the kth subinterval is given by $x_{k-1} \le x \le x_k$. Note $\Delta x = x_k - x_{k-1}$. This subinterval is the short side of a long thin rectangle. Let $f(x) = 6x - x^2$. Let x_k^* denote the midpoint of the kth subinterval. This means $x_k^* = (1/2)(x_{k-1} + x_k)$. Make the length of the kth rectangle equal to $f(x_k^*)$.

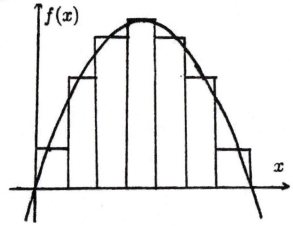

When the kth thin rectangle is revolved about the x axis the solid generated is a short cylinder with radius $f(x_k^*)$ and height Δx. We will revolve all these thin rectangles about the x axis. This gives us a collection of short cylinders that are beside each other.

Let us now turn our attention to the given solid of revolution. As stated above we slice the given solid S into slices like a loaf of bread. The kth slice of S is obtained by revolving that section of the region R which is such that $x_{k-1} \le x \le x_k$ about the x axis.

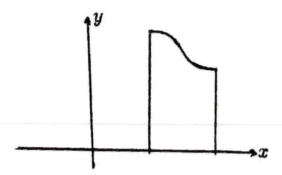

Let us compare the volume of the kth slice of the solid of revolution S with the volume of the kth small cylinder. A look at the cross section of these two solids in the xy plane makes it clear that the two volumes are approximately equal.

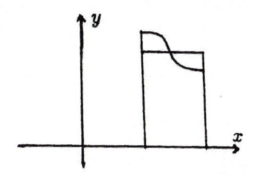

The volume of the kth short cylinder is approximately equal to the volume of the kth slice of the solid S. We already discussed how to find the volume of the kth short cylinder. The volume of the kth short cylinder is $\pi(\text{radius})^2(\text{height}) = \pi[f(x_k^*)]^2(\Delta x)$. The sum of the volumes of all the short cylinders is

$$\sum \pi[f(x_k^*)]^2[\Delta x].$$

The sum of the volumes of all the short cylinders is a good approximation to the volume of the solid of revolution. The volume of the solid S is approximately equal to

$$\sum_{k=1}^{n} \pi[f(x_k^*)]^2[\Delta x] = \sum_{k=1}^{n} \pi[6x_k^* - (x_k^*)^2]^2[\Delta x].$$

The more subdivisions we use on the x axis the closer the sum of the volumes of the short cylinders is to the actual volume of the solid S. The number of

short cylinders is n. The limit as n approaches infinity is the actual volume of S.

$$\text{volume of } S = \lim_{n \to \infty} \sum_{k=1}^{n} \pi [6x_k^* - (x_k^*)^2]^2 \Delta x$$

$$= \lim_{\Delta x \to 0} \sum_{k=1}^{n} \pi [6x_k^* - (x_k^*)^2]^2 \Delta x.$$

Note that in the notation used to define a definite integral $\Delta x = \frac{b-a}{n}$ and $x_k^* = a + \frac{2k-1}{2n}(b-a)$, the midpoint of the kth subinterval. With this in mind, we recognize this limit as the definite integral of $\pi(6x - x^2)^2$ on the interval $0 \leq x \leq 6$. Therefore,

$$\text{Volume of } S = \int_0^6 \pi [6x - x^2]^2 dx$$

$$= \pi \int_0^6 (36x^2 - 12x^3 + x^4) dx$$

We then evaluate this definite integral.

$$\text{Volume of } S = \pi \left[12x^3 - 3x^4 + (x^5/5) \Big|_0^6 \right] = \frac{1296}{5} \pi.$$

Example 2. Let R denote the region bounded by the parabola $y = 9 - x^2$ and the line $y = -x + 3$ and is on the positive side of the vertical line $x = -1$. Find the volume of the solid generated when the region R is revolved about the x axis.

Solution. In trying to find the volume of a solid of revolution we can use some of the same ideas we used in finding area. In Example 1 we saw the advantage of finding the volume of a region that is resting on the x axis, that is, a region whose lower boundary is the x axis.

For what values of x do the curves that are the graphs of $y = 9 - x^2$ and $y = -x + 3$ intersect? In order to find these values we solve the equations $y = 9 - x^2$ and $y = -x + 3$ simultaneously.

$$9 - x^2 = -x + 3$$

$$0 = x^2 - x - 6$$

$$(x - 3)(x + 2) = 0$$

$$x = 3 \text{ or } x = -2.$$

The curves intersect at the point where $x = -2$ and at the point where $x = 3$. This shows that the vertical line which is the boundary of R on the right is $x = 3$. The y coordinate when $x = 3$ is $y = 0$. We can find the y coordinate when $x = -2$. It is $y = -(-2) + 3 = 5$. The point of intersection is $(-2, 5)$. The vertical line which is the boundary of the region on the left is $x = -1$. This is helpful in making a sketch of R.

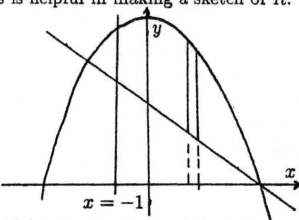

We divide the interval $-1 \le x \le 3$ on the x axis into subintervals of length Δx. We construct long thin rectangles parallel to the y axis of width Δx. The top end of the rectangle is on the parabola $y = 9 - x^2$ and the bottom end of the rectangle is on the line $y = -x + 3$. In order to be more exact let x_k^* denote the midpoint of the kth subinterval on the x axis. Let $f(x) = 9 - x^2$ and $g(x) = -x + 3$. The y coordinate of the top edge of the thin rectangle is $f(x_k^*) = 9 - (x_k^*)^2$ and the y coordinate of the bottom edge of the rectangle is $g(x_k^*) = -x_k^* + 3$. The length of the long side of the rectangle is

$$\text{length} = f(x_k^*) - g(x_k^*) = [9 - (x_k^*)^2] - [-x_k^* + 3].$$

When one of these long thin rectangles is revolved about the x axis it generates a solid which is a short cylinder with a hole in the middle. This solid is shaped like a washer. The volume of the washer is the volume of the complete short cylinder minus the volume of the hole. The volume of a short cylinder is $(\pi)(\text{radius})^2(\text{thickness})$. Let ΔV denote the volume of the washer, then

$$\Delta V = \pi[\text{outer radius}]^2[\text{thickness}] - \pi[\text{inner radius}]^2[\text{thickness}]$$
$$= \pi[f(x_k^*)]^2[\Delta x] - \pi[g(x_k^*)]^2[\Delta x]$$
$$= \pi[9 - (x_k^*)^2]^2[\Delta x] - \pi[-x_k^* + 3]^2[\Delta x]$$

The sum of the volumes of all washers is

$$\sum \pi[9 - (x_k^*)^2]^2[\Delta x] - \pi[-x_k^* + 3]^2[\Delta x].$$

The volume of S is equal to the limit of this sum as the number of subdivisions of the interval $-1 \le x \le 3$ becomes infinite. This limit is the following definite integral.

$$\text{Volume of } S = \pi \int_{-1}^{3} (9 - x^2)^2 dx - \pi \int_{-1}^{3} (-x + 3)^2 dx.$$

$$= \pi \int_{-1}^{3} [(9 - x^2)^2 - (-x + 3)^2] dx.$$

Squaring we get

$$\pi \int_{-1}^{3} [x^4 - 19x^2 + 6x + 72] dx = \pi \left[\frac{x^5}{5} - \frac{19}{3}x^3 + 3x^2 + 72x \Big|_{-1}^{3} \right] = \frac{2752}{15}\pi.$$

If the units on the x axis are meters, then the units of volume are cubic meters.

The short cylinders used in this method of finding the volume of a solid are often called disks. For that reason this method is usually referred to as "the method of disks".

In writing up this problem as a homework problem all the details about dividing the interval on the x axis into subintervals and so forth should go through a person's mind. But the write up only needs to include the essentials. Every problem should start with a sketch of the region R in the xy plane. Let ΔV denote the volume of the typical short cylinder or in this case the volume of the cylinder with the longer radius $9 - x^2$ minus the volume of the smaller cylinder with the shorter radius $-x + 3$.

$$\Delta V = [\text{outer radius}]^2 \Delta x - \pi[\text{inner radius}]^2 \Delta x$$
$$= \pi[9 - x^2]^2 \Delta x - \pi[-x + 3]^2 \Delta x$$
$$= \pi[(9 - x^2)^2 - (-x + 3)^2] \Delta x.$$

Add together all these elements of volumes, take the limit as the number approaches infinity, and we get that the volume of S is given by the definite integral

$$\text{volume of } S = \pi \int_{-1}^{3} [(9 - x^2)^2 - (-x + 3)^2] dx.$$

Then evaluate the integral.

Exercises

1. Let R denote the region bounded above by the curve whose equation is $y = 10x - x^2$ on the right by the line $x = 6$ and below by the x axis. Find the volume of the solid S generated when the region R is revolved about the x axis.

2. Let R denote the region bounded by the curve $y = \sqrt{9 - x^2}$ and the x axis. Find the volume of the solid S generated when the region R is revolved about the x axis.

3. Let R denote the region bounded by the parabola $y = 20 - x^2$ and the parabola $y = x^2 + 2$. Set up the definite integral which when evaluated will give the volume of the solid generated when the region R is revolved about the x axis.

4. Let R denote the region bounded by the parabola $y = x^2 + 2$ and the line $y = x + 4$. Set up the definite integral which when evaluated will give the volume of the solid generated when the region R is revolved about the x axis.

5. Let R denote the region bounded by the sine curve $y = \sin x$ and the x axis for $0 \leq x \leq \pi$. Find the volume of the solid generated when the region R is revolved about the x axis.

6405 Volume by Cylindrical Shells

We have another method for finding the volume of a solid of revolution. In this method we slice the solid in a different way. We do not slice it like a loaf of bread. One of the advantages of this method is that it usually results in integrals which are easier to evaluate.

We begin by looking at a fundamental element of volume which we get when we use the new method of slicing. Suppose we have a long thin rectangle R with one of the short sides of the rectangle a line segment on the x axis. Let w denote the width of this rectangle and h denote the height. Let x denote the distance between the centerline of this rectangle and the y axis.

Let S denote the solid of revolution generated by revolving this rectangle about the y axis. The solid S looks like a tin can with no top and no bottom. The solid S is just the metal used to construct the can and not what is inside the can. We would like to know the volume of the solid S. (We could find the volume of S using differentials.) Let us suppose we cut the solid S (the can) along a vertical line which is perpendicular to the base and flatten out the result to form a rectangle. The volume of the solid S is then equal to the length of one side of this rectangle times the length of the other side times the thickness. One side of the rectangle is h, the height of the can. The length of the rectangle is equal to the circumference of the can which is 2π (radius) $= 2\pi x$. The thickness is equal to w.

$$\text{Volume of } S = (\text{height of can})(\text{circumference})(\text{thickness})$$
$$= (h)(2\pi x)w.$$

In the following discussion we will use $w = \Delta x$ and $h = y$ so that the volume of such a can is given by volume $= 2\pi xy(\Delta x)$.

Example 1. Let R denote the region bounded by the graph of $y = 6x - x^2$ and the x axis. Let S denote the solid generated by revolving the region R about the y axis. Find the volume of the solid S.

Solution. The solid S in this example looks a lot like a doughnut that is flat on the bottom. Also there is really not a hole in the middle. The y axis is an axis of symmetry for the solid S. The curve $y = 6x - x^2$ crosses the x axis at the points where $x = 0$ and $x = 6$. The section of curve bounding the region R is the graph of $y = 6x - x^2$ for $0 \leq x \leq 6$. Therefore, divide the interval $0 \leq x \leq 6$ into subintervals of length Δx. Let x_k^* denote the midpoint of the kth subinterval. Let $f(x) = 6x - x^2$, then $f(x_k^*) = 6x_k^* - (x_k^*)^2$. We draw long thin rectangles in the region R. The base of the kth rectangle is the kth subinterval and the height of the kth rectangle is $f(x_k^*)$.

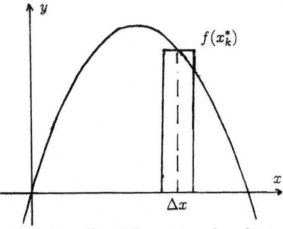

The solid generated by revolving the kth rectangle about the y axis is shaped like a tin can open at the top and bottom. This is a case like the one discussed above. We already found the volume of such a solid. It is the material required to construct the can. The volume is

$$2\pi x f(x_k^*)\Delta x,$$

Let us fill the figure in the xy plane with rectangles. When we revolve all these rectangles about the y axis the solids generated are a collection of cans one inside the other. The sum of all these cans is approximately equal to the solid S.

The solid S is approximated by a collection of cans one inside the other. The sum of the volumes of all these cans is

$$\sum_{k=1}^{n} 2\pi x f(x_k^*)\Delta x.$$

The volume of the solid S is the limit of this sum as the number of rectangles (tin cans) becomes infinite. This limit is also a definite integral. Therefore, the volume of S is

$$\text{volume of } S = \int_0^6 2\pi x f(x)dx = 2\pi \int_0^6 x[6x - x^2]dx.$$

Evaluating the integral

$$2\pi \int_0^6 (6x^2 - x^3)dx = 2\pi[2(6)^3 - (1/4)(6)^4]$$
$$= 2\pi(6^3)(1/2) = \pi(6)^3.$$

Example 2. Let R denote the region bounded above by the parabola $y = 6x - x^2$, below by the line $y = -2x$, and on the left by the line $x = 2$. Let S denote the solid generated by revolving the region R about the y axis. Find the volume of S.

Solution. For what values of x do the curves $y = 6x - x^2$ and $y = -2x$ intersect? Solve the equations $y = 6x - x^2$ and $y = -2x$ simultaneously.

$$6x - x^2 = -2x$$
$$x^2 - 8x = 0$$
$$x = 0 \text{ or } x = 8.$$

When making a sketch of R it is helpful to find the corresponding values of y. When $x = 0$, $y = 0$. When $x = 8$, $y = -16$. A sketch of R.

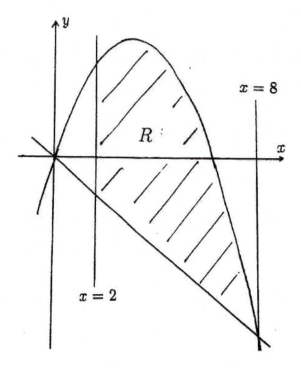

The solid S is shaped like an odd doughnut which does have a hole in the middle of radius 2. The region R in the xy plane is bounded on the left by the vertical line $x = 2$ and on the right by the vertical line $x = 8$. Let $f(x) = 6x - x^2$ and $g(x) = -2x$. Divide the interval $2 \leq x \leq 8$ on the x axis into short subintervals of length Δx. Draw long thin rectangles of width Δx with the rectangles starting on the bottom curve $y = -2x$ and extending to the top curve $y = 6x - x^2$. What is the length of a typical long thin rectangle? Suppose the rectangle is in the space where $2 \leq x \leq 6$.

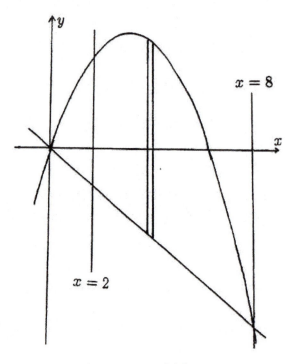

Let x_k^* denote the midpoint of the kth subinterval on the x axis. The distance from the x axis to the top curve is $f(x_k^*) = 6x_k^* - (x_k^*)^2$. The distance from the x axis to the bottom curve is $(-1)g(x_k^*) = (-1)[-2x_k^*] = 2x_k^*$. Since x is positive, the expression $-2x$ is negative. Distance (length) is positive. We must use $(-1)(-2x)$ to get length. The length of the kth rectangle is

$$f(x_k^*) + (-1)g(x_k^*) = [6x_k^* - (x_k^*)^2] - [-2x_k^*].$$

Suppose we look at a typical rectangle constructed when $6 \leq x \leq 8$.

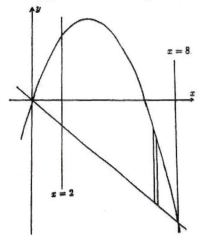

What is the length (height) of this rectangle? As already noted the distance from the x axis to the line $y = g(x) = -2x$ is given by $(-1)g(x_k^*) = 2x_k^*$. On this section of the parabola $y = 6x - x^2$ the y coordinate is negative. It follows that the distance from the x axis to the curve $y = f(x) = 6x - x^2$ is $(-1)f(x_k^*) = (-1)[6x_k^* - (x_k^*)^2]$. The sketch makes it clear that the length of the rectangle is given by distance from x axis to the line subtract the distance from the x axis to the parabola.

$$\text{length of rectangle} = (-1)g(x_k^*) - [(-1)f(x_k^*)] = f(x_k^*) - g(x_k^*).$$

Thus we see that no matter where the little rectangle is located

$$\text{length of rectangle} = f(x_k^*) - g(x_k^*).$$

Note that as we work more problems we see a general rule. The length of the thin rectangle equals the y coordinate of the top curve minus the y coordinate of the bottom curve. We also saw this when finding area. The

317

solid generated when this large thin rectangle is rotated about the y axis is a tin can. The volume of this can is

$$\text{Volume} = 2\pi(\text{radius})(\text{height})(\text{thickness})$$
$$= 2\pi(x_k^*)[f(x_k^*) - g(x_k^*)]\Delta x$$

The sum of the volume of all cans generated by all the rectangles is

$$\sum 2\pi(x_k^*)[f(x_k^*) - g(x_k^*)]\Delta x.$$

The limit of this sum as the number of subintervals becomes infinite is a definite integral which is the volume of the solid S. The volume of the solid S is

$$\int_2^8 2\pi x[f(x) - g(x)]dx = \int_2^8 2\pi x[(6x - x^2) - (-2x)]dx$$

$$= 2\pi \int_2^8 (8x^2 - x^3)dx = 648\pi$$

Example 3. Let R denote the region bounded by the line $y = x + 3$ and the parabola $y = x^2 - 4x + 3$. Set up the integral which when evaluated will equal the volume of the solid generated when the region R is revolved about the y axis.

Solution. We need to make a sketch of the region R. Solve the equations $y = x + 3$ and $y = x^2 - 4x + 3$ simultaneously

$$x + 3 = x^2 - 4x + 3$$
$$x^2 - 5x = 0$$
$$x = 0 \text{ and } x = 5.$$

When $x = 0$, we get $y = 3$. When $x = 5$, we get $y = 8$. The two curves which are the graphs of $y = x + 3$ and $y = x^2 - 4x + 3$ intersect at the points $(0,3)$ and $(5,8)$. We plot these points first in order to make a better sketch. Also note that $x^2 - 4x + 3 = (x - 3)(x - 1)$. From this we see that the parabola $y = x^2 - 4x + 3$ crosses the x axis at $x = 1$ and $x = 3$.

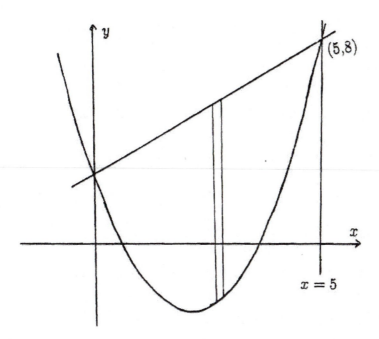

The region R is between the vertical lines $x = 0$ and $x = 5$. A typical rectangle is shown in the sketch. The volume of the solid generated when this rectangle is revolved about the y axis is

$$\Delta V = 2\pi[\text{radius}][\text{length of rectangle}][\text{thickness}].$$

The length of the rectangle equals the y coordinate of the line subtract the y coordinate of the parabola. The volume of the solid generated when this particular rectangle is revolved about the y axis is

$$\Delta V = 2\pi[x][(x + 3) - (x^2 - 4x + 3)]\Delta x.$$

We approximate the volume of the solid S by the sum of the volume of all such cans. Taking the limit as Δx approaches zero of the sum we get that the volume of the solid generated by revolving R about the y axis is given by the definite integral

$$\text{volume of } S = 2\pi \int_0^5 (x)[(x + 3) - (x^2 - 4x + 3)]dx.$$

Let us say a few words about notation when doing practical problems whose solution involves integrals. We have been using x and y as our variables.

319

We could, of course, choose any other letters. In setting up a practical problem we usually start with one or more equations. But in order to use integrals to solve the problem we must find a function to integrate. This function must be defined explicitly. In our study of calculus we never try to integrate a function which is defined implicitly using an equation. For example, suppose we say that y is defined as a function of x by the equation $y^3 + 4x^2y = 5$. We never try to integrate such a function. We need an explicit function $f(x)$. We often need to graph this function which may be described as graphing the equation $y = f(x)$. We almost always set up problems so that the independent variable is called x or the variable of integration is x. We graph x on the horizontal axis in the plane. This helps in communicating the ideas as we are all more familiar with this situation.

When the independent variable is time we usually call it t. When the independent variable is height we often decide to call it y as this seems to be in keeping with past practice that the vertical coordinate is y. When working problems a person may choose any letter he wishes for the independent variable. People often choose the first letter of some noun involved in the problem.

No matter what letter is used for the independent variable we will always construct the graph of a function with the independent variable on the horizontal axis. When setting up a practical problem you need to be aware of the role played by the independent variable in its solution. In some calculus texts you will see the independent variables for some of these volume problems graphed on the vertical axis. If you look at these books, you need to take this into account.

Exercises

1. Let R denote the region bounded by the parabola $y = 10x - 2x^2$ and the x axis and which is to the right of the line $x = 1$. Let S denote the solid generated when the region R is revolved about the y axis. Find the volume of S.

2. Let R denote the region bounded by $y = 8x - x^2$ and $y = -x$. Sketch a graph of R. Draw a typical long thin rectangle. Find the usual expression for the width and height of this rectangle.

3. Let R denote the region bounded by the curve $y = (x^2 + 1)^{-1}$, the y

320

axis, the x axis, and the line $x = 4$. Let S denote the solid generated when the region R is revolved about the y axis. Find the volume of S.

4. Let R denote the region bounded by the parabola $y = -x^2 + 6x - 5$ and the line $y = -x + 1$. Let S denote the solid generated when the region R is revolved about the y axis. Set up the definite integral which when evaluated will equal the volume of S.

5. Let R denote the region bounded by the parabola $y = -x^2 + 6x - 5$ and the parabola $y = x^2 - 4x + 3$. Let S denote the solid generated when the region R is revolved about the y axis. Set up the definite integral which when evaluated will equal the volume of S.

6407 Hard Volumes of Revolution

In our previous work we have found the volumes of solids of revolution in the case when the region R was revolved about the x-axis and in the case when the region R was revolved about the y axis. We are now going to look at solids of revolution formed when the region R is revolved about lines which are parallel to the x axis and where the region R is revolved about lines which are parallel to the y axis.

Example 1. Let R denote the region bounded by the parabola $y = x^2 + 1$ and the line $y = 5$. Let S denote the solid generated when R is revolved about the line $y = -3$. Find the volume of S.

Solution. The parabola $y = x^2 + 1$ intersects the line $y = 5$ where $x^2 + 1 = 5$ or $x = -2$ and $x = 2$.

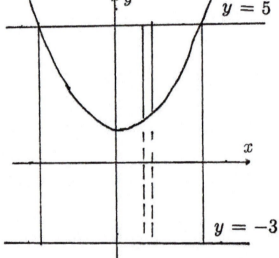

Divide the interval $-2 \le x \le 2$ into subintervals of length Δx. Draw a long thin rectangle of width Δx with its bottom on the parabola $y = x^2 + 1$ and its top end on the line $y = 5$. If we revolve this rectangle about the line $y = -3$ the solid generated is a short cylinder of height Δx with a hole in the middle. The generated solid could be described as a washer. The volume of a washer with a hole in the middle is the volume of the complete cylinder subtract the volume of the hole. The volume of a short cylinder is $\pi(\text{radius})^2(\Delta x)$. The outer radius of the washer is the distance from the line $y = -3$ to the line $y = 5$. This distance is the same for all values of x and is equal to 8. The inner radius of the washer is the distance from the line $y = -3$ to the parabola $y = x^2 + 1$. This is the distance from the x axis to the line $y = -3$, which is 3, added to the distance from the x axis to the parabola, which is the y coordinate of the points on the x axis. This distance is $3 + y = 3 + (x^2 + 1) = x^2 + 4$. The volume of the washer is

$$\Delta V = \pi[\text{outer radius}]^2(\Delta x) - \pi[\text{inner radius}]^2(\Delta x)$$
$$= \pi[8]^2 \Delta x - \pi[x^2 + 4]^2 \Delta x$$
$$= \pi[(8)^2 - (x^2 + 4)^2]\Delta x.$$

Adding the volumes of all washers and taking the limit as Δx approaches zero we get that the volume of the solid S is given by the definite integral

$$\pi \int_{-2}^{2} [64 - (x^2 + 4)^2]dx = \pi \int_{-2}^{2} [48 - 8x^2 - x^4]dx$$
$$= \pi[48(2 + 2) - \frac{8}{3}(8 + 8) - \frac{1}{5}(32 + 32)] = \frac{2048}{15}\pi.$$

Example 2. Let R denote the region bounded by the parabola $y = x^2 - 4x$ and the line $y = x$. Let S denote the solid generated when the region R is revolved about the line $y = 5$. Set up the integral which when evaluated will equal to the volume of the solid S.

Solution. The parabola $y = x^2 - 4x$ intersects the line $y = x$ where $x^2 - 4x = x$ or where $x = 0$ and $x = 5$. When $x = 0$, we get $y = 0$. When $x = 5$, we get $y = 5$. The parabola and the line intersect at $(0,0)$ and $(5,5)$.

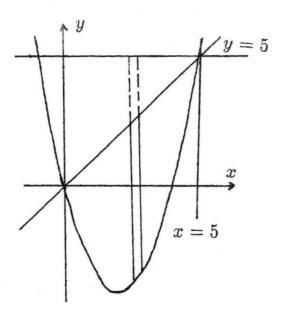

The line $y = 5$ only touches the region R. The figure is between the vertical lines $x = 0$ and $x = 5$. We divide the interval $0 \leq x \leq 5$ into subintervals of length Δx. Draw a typical rectangle of width Δx. The bottom end of the rectangle is on the parabola $y = x^2 - 4x$ and the top end is on the line $y = x$. When this rectangle is revolved about the line $y = 5$ a solid in the shape of a wisher is generated. The washer is a short cylinder with a hole in the middle. The inner radius of the washer is the distance from the line $y = 5$ to the line $y = x$. This distance is $5 - x$. The outer radius of the washer is the distance from the line $y = 5$ to the parabola $y = x^2 - 4x$. Let us look at the case when Δx is in the interval $0 \leq x \leq 4$. In this case the parabola is below the x axis. The value of y is negative. The distance from the x axis to the parabola is $[-y]$. The distance from the line $y = 5$ to the parabola is $5 + [-y] = 5 - y$. If the long thin rectangle has its lower end on the section of parabola in the interval $4 \leq x \leq 5$, then we easily see that the length of the rectangle is $5 - y$. Since y denotes the y coordinate of points on the parabola, the outer radius of the washer is $5 - y = 5 - (x^2 - 4x)$. Let ΔV denote the volume of the washer, then

$$\Delta V = \pi[\text{outer radius}]^2 (\Delta x) - \pi[\text{inner radius}]^2 (\Delta x)$$
$$= \pi[5 - (x^2 - 4x)]^2 \Delta x - \pi[5 - x]^2 (\Delta x)$$
$$= \pi[(-x^2 + 4x + 5)^2 - (5 - x)^2]\Delta x.$$

Adding together the volumes of all the washers and taking the limit as Δx approaches zero, we get that the volume of the solid S is given by the definite integral

$$\pi \int_0^5 [(-x^2 + 4x + 5)^2 - (5 - x)^2]dx$$

$$= \pi \int_0^5 [x^4 - 8x^3 + 5x^2 + 50x]dx.$$

Example 3. Let R denote the region bounded by the parabola $y = 9 - x^2$ and the x axis. Let S denote the solid generated when the region R is revolved about the line $x = -5$. Find the volume of the solid S.

Solution. The parabola $y = 9 - x^2$ crosses the x axis at the points where $x = -3$ and $x = 3$

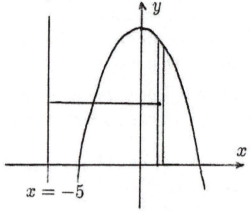

We divide the interval $-3 \le x \le 3$ into subintervals of length $\Delta x = 6/n$. Draw a typical long thin rectangle of width Δx. The top end of the rectangle is on the parabola $y = 9 - x^2$ and the bottom end is on the x axis. When this rectangle is revolved about the line $x = -5$ a solid in the shape of a tin can is generated. The height of the can is $9 - x^2$. When the long thin rectangle is on the positive side of the y axis, then it is easy to see that the radius of the can is $5 + x$. Suppose the long thin rectangle is on the negative side of the y axis. When this is the case the value of x is negative, that is, $x < 0$. The distance from the y axis to the rectangle is $[-x]$. The radius of the can is $5 - [-x] = 5 + x$. No matter where the rectangle is located the distance from the rectangle to the line $y = -5$ is $5 + x$.

Let ΔV denote the volume of a single can

$$\Delta V = 2\pi(\text{radius})(\text{height})(\text{thickness})$$
$$= 2\pi(5 + x)(9 - x^2)(\Delta x)$$

Adding together the volumes of all the tin cans and taking the limit as n approaches infinity, we get that the volume of the solid S is given by the following definite integral.

$$\text{Volume} = \int_{-3}^{3} 2\pi(5 + x)(9 - x^2)dx$$

$$= 2\pi \int_{-3}^{3} (-x^3 - 5x^2 + 9x + 45)dx$$

$$= 2\pi[(-1/4)(81 - 81) - (5/3)(27 + 27) + (9/2)(9 - 9) + 45(3 + 3)]$$

$$= 360\pi.$$

Example 4. Let R denote the region bounded by the parabola $y = x^2 + 1$ and the line $y = 2x + 4$. Let S denote the solid generated when the region R is revolved about the line $x = 4$. Set up the integral which when evaluated will be equal to the volume of the solid S.

Solution. In order to find the points where the parabola and the line intersect we solve the equations $y = x^2 + 1$ and $y = 2x + 4$ simultaneously.

$$x^2 + 1 = 2x + 4$$
$$x^2 - 2x - 3 = 0$$
$$(x - 3)(x + 1) = 0$$
$$x = -1 \text{ or } x = 3.$$

The y coordinate corresponding to $x = -1$ is $y = 2$. The y coordinate corresponding to $x = 3$, is $y = 10$. The parabola and the line intersect at $(-1, 2)$ and $(3, 10)$. Plot the points $(-1, 2)$ and $(3, 10)$ and sketch R.

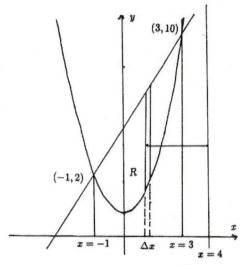

The region R is bounded by the vertical lines $x = -1$ and $x = 3$. We divide the interval $-1 \leq x \leq 3$ into subintervals of length $\Delta x = 4/n$. Draw a typical long thin rectangle of width Δx. The top end of the rectangle is on the line $y = 2x + 4$. The bottom end of the rectangle is on the parabola $y = x^2 + 1$. When this rectangle is revolved about the line $x = 4$ the solid generated is in the shape of a tin can. The height of the can is

$$y_{\text{line}} - y_{\text{parabola}} = [(2x + 4) - (x^2 + 1)] = [-x^2 + 2x + 3].$$

When the long thin rectangle is on the positive side of the y axis, then it is easy to see that the radius of the can is $4 - x$. When the thin rectangle is on

the negative side of the y axis, then the values of x inside the rectangle are negative, that is, $x < 0$. Distance is positive. The distance from the y axis to the thin rectangle is $[-x]$. The radius of the can is again $4 + (-x) = 4 - x$.

Let ΔV denote the volume of the can, then

$$\Delta V = 2\pi(\text{radius})(\text{height})(\text{thickness})$$
$$= 2\pi(4 - x)(-x^2 + 2x + 3)(\Delta x).$$

Adding together the volumes of all the cans and taking the limit as $\Delta x \to 0$, we get that the volume of the solid S is given by

$$\text{Volume} = \int_{-1}^{3} 2\pi(4 - x)(-x^2 + 2x + 3)dx$$
$$= 2\pi \int_{-1}^{3} (x^3 - 6x^2 + 5x + 12)dx.$$

Exercises

1. Let R denote the region bounded by the parabola $y = 4 - x^2$ and the x axis. Let S denote the solid generated when R is revolved about the line $y = -3$. Find the volume of S.

2. Let R denote the region bounded by the parabola $y = x^2 - 3x$ and the line $y = x$. Let S denote the solid generated when R is revolved about the line $y = -4$. Set up the integral which when evaluated will be equal to the volume of S.

3. Let R denote the region bounded by the parabola $y = 4x - x^2$ and the x axis. Let S denote the solid generated when R is revolved about the line $x = -3$. Find the volume of S.

4. Let R denote the region bounded by the parabola $y = x^2 - 3x$ and the line $y = x$. Let S denote the solid generated when R is revolved about the line $x = -2$. Find the volume of S.

5. Let R denote the region bounded by the parabola $y = x^2 - 3x$ and the line $y = x$. Let S denote the solid generated when R is revolved about the line $x = 5$. Set up the integral which when evaluated will be equal to the volume of S.

6. Let R denote the region bounded by $y = 4x - x^2$ and $y = -x$. Let S denote the solid generated when R is revolved about the line $x = 6$. Set up the integral which when evaluated will equal to the volume of S.

7. Let R denote the region bounded by the parabola $y = 4x - x^2$ and the line $y = -2x$. Let S denote the solid generated when R is revolved about the line $y = 4$. Set up the integral which when evaluated will be equal to the volume of S.

8. Let R denote the region bounded by the parabola $y = -x^2 + 5x - 4$ and the line $y = -x + 1$. Let S denote the solid generated when R is revolved about the line $y = 3$. Set up the integral which when evaluated will equal to the volume of S.

6411 Work

A vector has both magnitude and direction. Force has both magnitude and direction. Gravity is a force. It has a certain magnitude depending on mass and its direction is toward the center of the earth. A force is represented using a vector. In three dimensional space indicating the direction of a vector is somewhat complicated. We are going to discuss situations where all the forces involved act along the same straight line. All displacements are also along this same line. When all vectors act along the same line we have simple method we use to denote the direction of the vectors. Since there are only two possible directions we can indicate of the direction of vectors by saying that one direction is the plus direction and the other direction is the minus direction. We can do this by just using the usual positive and negative signs. The magnitude of the vector is still the magnitude of the vector and is a real number (a scalar).

We denote the vector of magnitude 15 and in the plus direction by $(+15)$ and the vector of magnitude 20 and in the negative direction by (-20). This means that we begin each problem by choosing a line of action and by choosing a positive direction.

Suppose a weight is suspended at the end of a rope and the weight is motionless. There are two forces acting on the weight. There is the force in the upward direction exerted by the rope. There is the force in the downward direction exerted by gravity. These two forces have the same magnitude but are in opposite directions.

We are going to discuss the work done by a force during a given displacement along a line. We will only consider the case when the direction of the force is parallel to the direction of the displacement. If the force has the same direction as the displacement, then

$$\text{work} = (\text{magnitude of force}) \times (\text{displacement})$$

If the force has the opposite direction from the displacement, then

$$\text{work} = (\text{minus})(\text{magnitude of force})(\text{displacement}).$$

Consider the case of a 25 lb weight hanging from a rope through a pulley attached at a high point. Suppose a person raises the weight 8 feet by pulling on the rope. If we choose up as positive, then the displacement is

(+8) feet. The force exerted on the weight by the rope (person) is (+25) lbs. The work done by the person is

$$(25)(8) = 200 \text{ ft} - \text{lbs}.$$

If we choose down as positive, then the displacement is (−8) feet. The force exerted on the weight by the rope is (−25) lbs. The work done by the person is

$$(-25)(-8) = 200 \text{ ft} - \text{lbs}.$$

There is another force acting on the weight. This is the force of gravity. The magnitude of the force of gravity is 25 lbs and the direction is down. Let us compute the work done by the force of gravity instead of the work done by the person. Suppose the weight is raised 8 feet. Let us choose up as positive. The displacement is (+8) feet. The force of gravity is (−25) lbs. The work done by gravity is

$$(-25)(8) = -200 \text{ ft} - \text{lbs}.$$

Suppose the weight is raised 8 feet and down is chosen as positive. The displacement is (−8) ft and the force of gravity is (+25) lbs. The work done by gravity is

$$(25)(-8) = -200 \text{ ft} - \text{lbs}.$$

Let us look at a different action. Suppose the person lowers the weight 8 feet. If we choose up as positive, then the displacement is (−8) feet. The force exerted by the rope (person) is (+25) lbs. The work done by the person is

$$(25)(-8) = -200 \text{ ft} - \text{lbs}.$$

Suppose the person lowers the weight 8 feet. If we choose down as positive, then the displacement is (+8) feet. The force exerted by the rope is (−25) lbs. The work down by the person is

$$(-25)(8) = -200 \text{ ft} - \text{lbs}.$$

Suppose the person lowers the weight 8 feet. Let us find the work done by the force of gravity. If we choose up as positive, then the displacement is

(-8) ft. The force exerted on the weight by gravity is (-25) lbs. The work done by the force of gravity is

$$\text{work} = (-8)(-25) = 200 \text{ ft} - \text{lbs}.$$

Suppose the weight is lowered 8 feet. Let us find the work done by the force of gravity. If we choose down as positive, then the displacement is $(+8)$ ft. The force exerted on the weight by gravity is (25) lbs. The work done by the force of gravity is

$$\text{work} = (8)(25) = 200 \text{ ft} - \text{lbs}.$$

Suppose we are required to find the work done by the person in the above problems then it is not correct to find the work done by the force of gravity. We would be using the wrong force vector. It is clear that every solution of a work problem must begin by stating which direction is selected as the positive direction. This determines the sign on all numbers. In physics the above problems would often be solved by computing the change in potential. The solution by change in potential requires us to consider both positive and negative work.

In problems we will be working with objects that have weight. The two forces acting on the object will be the force exerted by a person and the force exerted by gravity. We will always be computing the work done by the person. We will always choose up as positive.

Example 1. Suppose a weight of 200 lbs is lifted up along a vertical line for 40 feet, find the amount of work done in lifting the weight.

Solution. Choose up as positive. The force is $+200$ lbs and the displacement is $+40$ feet
$$\text{work} = (200)(40) = 8,000 \text{ ft} - \text{lbs}.$$

In the engineering system of units the unit of work is ft-lbs. In the mks system of units the unit of work is Newton-meters. When working in the mks system of units it is important to remember that the unit of weight is Newtons not kilograms.

Problems such as Example 1 are obviously easy to solve. Also they do not involve calculus. In order to involve calculus either the force or the

displacement must be variable. Variable displacement is somewhat more difficult. In order to involve calculus we begin by using a variable force.

Suppose we have an object displaced along a line from $x = a$ to $x = b$ with $a < b$. Suppose a variable force $f(x)$ acts on the object. Note that since $a < b$ we have selected a positive direction. The units of distance can be either feet or meters. The units of force are either pounds or newtons. We begin by approximating the work required to move the object. Divide the number line from $x = a$ to $x = b$ into n equal length subintervals of length $\Delta x = \frac{b-a}{n}$. Denote the subdivision points by $x_0 = a$, $x_1, x_2, \ldots x_n = b$.

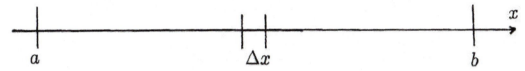

Let us approximate the work done to move the object through the kth subinterval. This is the part of the line such that $x_{k-1} \leq x \leq x_k$. Since the subinterval of length Δx is very short, we may assume that the force acting on the object is a constant over this subinterval. Let x_k^* denote the midpoint of the kth subinterval. Assume that the force has the constant value $f(x_k^*)$ for all x such that $x_{k-1} \leq x \leq x_k$. The work required to move the object through this subinterval is approximately $f(x_k^*)\Delta x$. Adding the work for all subintervals together we get the sum

$$\sum_{k=1}^{n} f(x_k^*)\Delta x.$$

This is approximately the work required to move the object from $x = a$ and $x = b$. Taking the limit as the number of subintervals becomes infinite we get the actual work. Recalling the definition of the definite integral we realize that this limit is an integral. This means that

$$\text{work} = \int_a^b f(x)dx.$$

Example 2. Suppose an object is moved along the x axis from $x = 0$ to $x = 10$ feet by a force of $(2x + 5)$ lbs. How much work is done by the force?

Solution. This is a variable force. Note that force is acting in the positive direction. We know the force as a function of the displacement coordinate, x. Therefore,

$$\text{work} = \int_0^{10} (2x + 5)dx = 150 \text{ ft}-\text{lbs.}$$

We are lucky in this example to have the expression for force given. We could have started this problem by requiring that the expression for force be found. We could have said that the weight was attached to a spring and was being pushed along a rough floor in a line, for example. In the future problems we will have to find the variable force. In most cases the force we will be considering is the force used to counteract the weight of an object.

Suppose we have a spring that resists both compression and stretching. Suppose one end of the spring is attached to a fixed wall and the other end is attached to a body of mass m. Assume that the body rests on a frictionless horizontal plane so that as we move the body back and forth on a line the spring compresses and stretches. Let x denote the distance of the body from its equilibrium position on the line. Equilibrium position is when the spring is neither stretched nor compressed. We use $x > 0$ to indicate the position when the spring is stretched and $x < 0$ to indicate position when the spring is compressed. The spring exerts a restoring force when either stretched or compressed.

Hooke's Law. Suppose one end of a spring is fixed and the other end is displaced by a distance x from its equilibrium position, then the force required to keep the spring displaced is

$$F_s = kx.$$

The constant k is called the spring constant. We are assuming that the spring is not stretched beyond its elastic limit. In problems we can be given the spring constant or we can be required to find the spring constant using the equation $F_s = kx$ from Hooke's Law. Hooke's Law is force equals spring constant times displacement.

Example 3. Suppose a body is attached to a very long spring as above and the spring has spring constant $k = 5$ lbs/ft. Find the work done in stretching the spring from 1 foot to 5 feet. The units of k are lbs/ft or Newtons/meter.

Solution. Since the spring constant is $k = 5$ the force is $F = 5x$.

$$\text{Work} = \int_1^4 (5x)\,dx = \frac{5}{2}x^2 \mid_1^4 = \frac{75}{2} \text{ ft lbs.}$$

Example 4. Suppose we have a very long spring attached to a wall. A force of 60 pounds stretches the spring 3 feet beyond its equilibrium length and exactly holds it in this position. Find the work done in stretching this spring from 1 foot to 5 feet beyond its rest length.

Solution. First, we need to find the spring constant. Use

$$\text{force} = (k)(\text{displacement})$$
$$60 = (k)(3)$$
$$k = 20 \text{ lbs/ft.}$$

We have force $F_s = 20x$. This is a variable force. The work is

$$\int_1^5 20x\,dx = 10x^2 \mid_1^5 = 240 \text{ ft lbs.}$$

Bucket and Chain Problems

Example 5. A chain hangs over a pulley from the top of a tall building. The chain is 200 feet long and weighs 4 lbs per foot. How much work is done to roll the first 120 feet of the chain over the pulley?

Solution. The chain moves up and down in a vertical line. We need a coordinate line on which to indicate the position of the lower end of the chain. Since the line is vertical let us call this coordinate y. We choose up as positive since this seems the natural thing. Choose $y = 0$ as the bottom end of the chain before the chain is lifted. The pulley is at $y = 200$. We are going to move the bottom end of the chain from $y = 0$ to $y = 120$. The force required to lift the chain is the weight of the chain that is still hanging down. The force that must be exerted to move the chain is an upward force and therefore positive. As the chain is being rolled up the pulley the force that must be exerted on the chain at the pulley is given by the weight of the part of the chain that is still hanging down. Suppose that the bottom end of the chain has already been moved up y feet, what is the force required to move the bottom end of the chain up to $y + \Delta y$ feet?

weight of chain = (linear density)(length of chain hanging down)

$$= (4 \text{ lb/ft})(200 - y) \text{ feet}$$
$$= 4(200 - y) \text{ lbs.}$$

The weight of the portion of the chain that is hanging down is $4(200 - y)$. The force that must be exerted to lift the chain is equal to the weight.

$$\text{work} = \int_0^{120} 4(200 - y)dy.$$
$$= 800y - 2y^2 \big|_0^{120}$$
$$= 67,200 \text{ ft} - \text{lbs.}$$

We can complicate the problem in Example 5 by adding a weight to the end of the chain. Placing a constant weight at the end of the chain does not really change how the problem is set up very much. The problem needs a variable weight.

Example 6. A 300 foot chain hangs over the edge of a tall building. The chain weighs 2 lbs per foot. There is a bucket of water at the end of the chain. Initially the bucket contains 1250 lbs of water (this is 20 ft^3). As the bucket is lifted, water leaks from the bucket at the rate of 4 lbs per foot that the bucket moves. How much work is done in lifting 200 feet of chain to the top of the building?

Solution. Note that water leaks from the bucket in a very special way. The amount of water that leaks out is a function of how far the bucket moves and is not a function of time. This method of leaking is used to make the "set-up" of the problem easier.

All the forces in this problem are the result of the weight of the objects involved. All forces act along the same vertical line. Since the line is vertical let us call this line the y axis. Let us do the "natural" thing and choose up as positive. The force which is doing the work acts in the upward direction and so is positive. The sign on all forces is plus. The magnitude of the forces are equal to the weight of the objects. Let $y = 0$ denote the initial position of the bucket. This makes $y = 300$ (feet) the top of the building. Suppose we have moved the bucket a distance y up the y axis, then the weight of the bucket is $1250 - 4y$. The length of the chain is $300 - y$. The weight of that part of the chain still hanging over the side of the building is $2(300 - y)$. The force required to lift the bucket plus chain from y to $y + \Delta y$ is

$$\text{force} = (1250 - 4y) + 2(300 - y)$$
$$= 1850 - 6y.$$

The bucket moves from $y = 0$ to $y = 200$. When the force is given as a function of displacement, then we have already shown that work is given by the integral

$$\text{work} = \int_0^{200} (1850 - 6y)dy$$

$$= 1850y - 3y^2 \Big|_0^{200} = 250,000 \text{ ft-lbs.}$$

Example 7. A 150 foot long chain hangs from the top of a tall tower. The chain weighs 5/2 lbs per foot. There is a large bucket at the end of the chain. Initially the bucket contains 2 cubic feet of water. As the bucket is lifted water drains **into** the bucket at the rate of 1/20 cubic foot for every foot the bucket is lifted. How much work is done in lifting the bucket the first 120 feet toward the top of the tower?

Solution. All forces act along a single vertical line. Denote this vertical line by y and choose up as the positive direction. Let $y = 0$ be the starting point of the bucket. This makes $y = 150$ the top of the tower. We need the weight of the water. The density of water is 62.5 lbs/ft^3. The magnitude of all forces is equal to the weight of the objects. The initial weight of the water is 125 lbs. Water is added to the bucket at the rate of $62.5/20 = 25/8$ lbs per foot. Suppose we have moved the bucket a distance y up the y axis, then the force required to keep the bucket moving has the same magnitude as the weight of chain hanging down plus the weight of water in the bucket. The length of chain hanging down is $150 - y$. The weight of chain hanging down is the length of the chain times the linear density of the chain.

$$\text{weight of chain} = (5/2)(150 - y).$$

The weight of the bucket is 125 lbs plus 25/8 lbs for every foot the bucket has moved. The bucket has moved y feet.

$$\text{weight of bucket} = 125 + (25/8)y.$$

The direction of the force doing the work is up and up is positive. This means that the force required to move the bracket is

$$\text{force} = (\text{plus sign})(\text{magnitude})$$
$$= (\text{plus sign})(\text{weight}).$$

The total force doing the work is

$$\text{force} = 5/2(150 - y) + [125 + (25/8)y]$$
$$= 500 + (5/8)y$$

The bucket is displaced from where $y = 0$ to where $y = 120$.

$$\text{work} = \int_0^{120} [500 + (5/8)y]dy$$
$$= 500y + (5/16)y^2\big|_0^{120} = 64,500 \text{ ft} - \text{lbs}.$$

In the MKS system the weight density of water is 9,800 newtons per cubic meter. An essential part of the solution of the bucket and chain problems is to state exactly what the variable of integration represents, state which direction is positive, and state what point on the displacement line is the zero point.

Before we look at the next problem: let us recall the definition of the definite integral $\int_a^b f(x)dx$. We divided $a \leq x \leq b$ into n short subintervals $x_i - x_{i-1} = \frac{b-a}{n} = \Delta x$. Note that $\Delta x > 0$. The integral was defined as

$$\lim_{n \to \infty} f(x_i^*)\Delta x = \int_a^b f(x)dx.$$

Note again that $b > a$. We used this definition when we developed facts about the definite integral. If we start with Δx negative, we get

$$\int_b^a f(x)dx = -\int_a^b f(x)dx.$$

When applying the definite integral to real world situations we need to remember these assumptions that were made in developing the definite integral, for one $\Delta x > 0$.

EXAMPLE 8. A bucket is attached to a chain and is being lowered from the top of a tall building. The linear density of the chain is 4 lbs/ft. The bucket plus water weighs 500 lbs when started at the top of the building. As the bucket is lowered water is added to the bucket at the rate of 6 lbs/ft.

The bucket is lowered by a person at the top of the building 250 feet to the ground. Find the work done by the person.

Solution. Choose up as positive. Let y denote the vertical coordinate with $y = 0$ at ground. Suppose the bucket has been lowered so that its height is y. What force does the person on the roof exert at this point? The distance from the roof to the bucket is $200 - y$.

Weight for bucket $= 500 + 6(250 - y)$
Weight for chain $= 4(250 - y)$
Total weight $= 500 + 1500 - 6y + 1000 - 4y = 3000 - 10y$

Note that as the person lowers the bucket he exerts a force equal to the weight in the positive direction which is up. We look at a small movement of the bucket. This movement is down which is in the negative direction and so the displacement is $(-\Delta y)$. Work is equal to force times displacement.

$$\Delta w = (3000 - 10y)(-\Delta y).$$

We take the limit as the number of subdivisions becomes infinite and we get

$$\text{Total work} = \int_0^{250} (3000 - 10y)(-dy)$$
$$= -3000y + 5y^2 \, \big|_0^{250}$$
$$= -3000(250) + 5(250)^2 = -437,500$$
$$\text{work} = -437,500 \text{ ft} - \text{lbs.}$$

As expected the value of work is negative because the potential energy decreased.

Some people might like to set up the problem with down as positive. In that case we would choose $y = 0$ at the roof. In this case

$$\text{force on bucket} = -(500 + 6y)$$
$$\text{force on chain} = -4y$$

$$\text{work} = \int_0^{250} [-(500 + 6y) - 4y]dy$$
$$= -\int_0^{250} (500 + 10y)dy = -437,500.$$

In doing these problems we have not considered some small forces involved in the motion. Suppose we are moving an object along a line. It must attain a non zero velocity. Some force (work) is needed to cause the weight to attain this velocity which means that it has some kinetic energy. Recall that (force)(distance = kinetic energy = $(1/2)mv^2$. We are assuming that the negative work which brings the object to a stop cancels the positive work which gave it the initial velocity. In any case we will ignore such forces whatever they are.

Exercises

1. An object is moved along the x axis from $x = 0$ to $x = 8$ feet by a force of $(3x^2 + 10)$ lbs. How much work is done by the force?

2. An object is moved along the x axis from $x = 2$ to $x = 10$ meters by a force of $(4x + 7)$ newtons. How much work is done by the force?

3. Suppose a body is attached to a very long spring which is attached to a wall. A force of 45 lbs stretches the spring 3 feet beyond its natural length and exactly holds it in that position. Find the work done in stretching the spring from 1 foot beyond its rest length to 5 feet beyond its rest length. Also find the work done in stretching this spring from 2 feet beyond its rest length to 6 feet beyond its rest length.

4. Suppose a very long spring is attached to a wall. A force of 52 lbs stretches the spring 5 feet beyond its natural length and exactly holds it in this position. Find the work done in stretching this spring from 1 foot beyond its equilibrium length to 4 feet beyond its equilibrium length. This is the work done by the displacing force which is the negative of the work done by the restoring force.

5. A chain hangs over a pulley from the top of a building 200 feet tall. The chain is 300 feet long so that 100 feet of the chain is on the ground. The chain weighs $(3/2)$ lbs per foot. How much work is required to roll up 100 feet of the chain over the pulley so that now the lower end of the chain just reaches the ground? Work is 30,000 ft-lbs.

6. A chain hangs over a pulley from the top of a tall building. The chain is 400 feet long, does not reach the ground, and weighs $5/2$ lbs per foot, that is, the linear density is $5/2$ lb/ft. If the linear density of a chain is ρ, then a section of the chain of length ℓ weighs $\rho\ell$. How much work is done to roll the first 250 feet of the chain over the pulley?

7. A 200 foot chain hangs over the edge of a tall building. The chain weighs 2.5 lb/ft. There is bucket of water at the end of the chain. Initially the bucket contains 800 lbs of water. As the bucket is lifted water leaks from the bucket at the rate of 3 lbs per foot. How much work is done in lifting the bucket the first 160 feet?

8. A chain 180 meters long hangs over the edge of a building that is 180 meters tall. The chain weighs 25 newtons per meter. There is a bucket at the end of the chain resting on the ground. Initially the bucket contains water weighing 4,000 newtons. As the bucket is lifted water leaks from the bucket at the rate of 20 newtons per meter that the bucket is lifted. How much work is done in lifting the bucket and chain from the point where the chain is 150 meters long to the point where the chain is 50 meters long?

9. A 200 foot chain hangs from the top of a tall tower. The chain weighs 13/4 pounds per foot. There is a large bucket at the end of the chain. Initially the bucket contains 20 ft^3 of water. As the bucket is lifted water drains into the bucket at the rate of 1/10 ft^3 per foot the bucket is lifted. How much work is done in lifting the bucket plus the chain from the point where 180 ft of chain hangs from the top to the point where 50 ft of chain hangs from the top. Work is 280,150 ft-lbs.

10. A 400 foot chain hangs over the edge of a very tall building. The weight density of the chain is 5/2 lb per foot. There is a bucket at the end of the chain that initially contains 200 lbs of water. As the bucket is lifted water is added into the bucket at the rate of 4 lbs per foot. How much work is done in lifting the bucket from where the chain is 300 feet long to where the chain is 50 feet long?

11. A dead pig is in a container on the top of a building 300 ft tall. The pig plus container weighs 500 lbs. The pig is lowered to the ground using a weightless rope. How much work is done by the person who lowered the pig? Note: Initial potential energy plus work equals final potential energy. You can **not** use this law to work the problem but you can use it to check your answer.

12. A chain with linear density of 5 lbs/ft is in a box on top of a tall building. The chain is attached to a bucket of water that weighs 2000 lbs. As the bucket is lowered using the chain water leaks from the bucket at the rate of 8 lbs/ft. The bucket is lowered down the side of the building using the chain 200 ft to the ground. One end of the chain is still at the top of

the building. How much work is done by the person who lowers the bucket and chain?

6413 Pumping Water

We are going to look at a different type of problem where we compute the work done. We are going to find the amount of work required to pump the water out of a tank. For this type of problem, we must use a method of solution which is slightly different from the method used on the bucket and chain problems. In these problems we usually do not find an expression for the force as a function of displacement. We use a different idea in order to find the integral for work.

Example 1. A large rectangular tank is full of water. The tank is 4 feet wide, 5 feet long, and 3 feet deep. Find the work required to pump all the water in the tank out an outlet 2 feet above the top of the tank. Use 62.5 lb/ft^3 as the weight density of water.

Solution. A sketch of the tank is helpful.

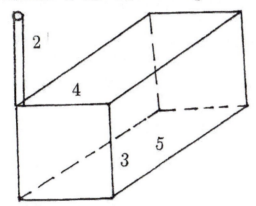

We must at least select a variable for the independent variable. In fact, let us draw a set of axis in the usual manner on the front side of the tank. Place the origin at the lower left corner with the y axis vertical and the x axis horizontal. A different placement of the coordinate axes will only change some of the details of the solution. Note that the water must be lifted up. The force that must be exerted on the water in order to lift it is in the upward direction. This is a good reason to choose up as positive. This keeps us from having to place minus signs on the forces. It is clear that the amount of work required to lift some water that is near the bottom of the tank out the outlet is greater than the work required to lift the same amount of water near the top of the tank out the outlet.

We want to find the work required to lift a small element of water out the outlet. We must select an element such that all the water in the element is the same distance below the outlet. The y axis for $0 \le y \le 3$ forms one edge of the tank. Divide this interval on the y axis into short subintervals each of length Δy. Consider a sheet of water parallel to the ground and of height Δy.

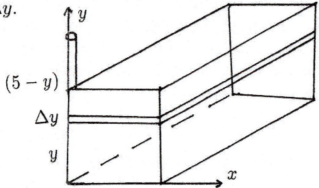

All the water in this sheet of water must be pumped the same distance in order to reach the outlet. Let ΔV denote the volume of the sheet of water and Δw the amount of work required to pump this sheet of water out the outlet. The distance from the origin to this sheet of water is y. The distance from the sheet of water to the outlet is $5 - y$. Let ρ denote the weight density of water, then the weight of water in the sheet is $\rho(\Delta V)$. The volume of the sheet is $(4)(5)(\Delta y)$. The weight of the sheet of water is

$$\rho(4)(5)(\Delta y).$$

Recall that work equals force times displacement.

$$\Delta w = (5 - y)(\rho)(4)(5)(\Delta y)$$
$$= (20\rho)(5 - y)\Delta y.$$

Find the work for each interval Δy, then the sum for all intervals is approximately the total work. If we take the limit as the number of subintervals becomes infinite, we get the actual work. This limit is given by a definite integral. Therefore,

$$\text{work} = \int_0^3 (20\rho)(5 - y)dy$$
$$= 210\rho.$$

Substituting $\rho = 62.5$, we get

$$\text{work} = 210(62.5) = 13,125 \text{ ft} - \text{lbs.}$$

Example 2. The cross section of a large tank is an isosceles triangle with one vertex on the ground. The edge of the triangle parallel to the ground is 12 feet long. The two sloping sides are each 10 feet long. The tank is 16 feet long and full of water. Find the work required to pump all the water to the level of the top of the tank.

Solution. We draw a set of coordinates on one end of the tank. We choose the y axis as vertical and the x axis as horizontal with the origin at the vertex which is on the ground. A sketch of the end of the tank is

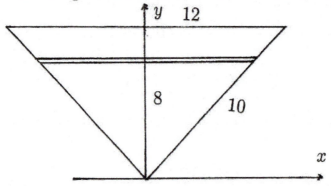

Divide the y axis for $0 \leq y \leq 8$ into subintervals of length Δy. We then have a rectangular sheet of water of thickness Δy parallel to the ground. Also all the water in this sheet is the same distance from the top of the tank. As we change y the width of the sheet of water changes. Call the width of the sheet of water $2x$. Let ΔV denote the volume of this sheet of water, then

$$\Delta V = (2x)(16)(\Delta y).$$

We are going to integrate with respect to y. This means that we need an expression for x in terms of y. It is fairly easy to find such an expression using similar triangles. However, we will find x as a function of y as follows. The side of the triangle is the first quadrant is part of a line. We will find the equation of that line. We know two points on this line. The points are $(0,0)$ and $(6,8)$.

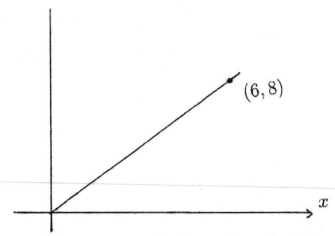

The general equation of a line through the point (x_1, y_1) with slope m is

$$y - y_1 = m(x - x_1).$$

Substituting $(x_1, y_1) = (0, 0)$, we get that the equation must be of the form

$$y = mx.$$

Substitution $(x, y) = (6, 8)$, we get $8 = m(6)$ or $m = 4/3$. The equation of the line is

$$y = (4/3)x \text{ or } x = (3/4)y.$$

The expression for ΔV is

$$\Delta V = (3/2)(y)(16)(\Delta y) = (24y)(\Delta y).$$

Multiplying by ρ, the density of water, we get that the weight of water in this element is $(\rho)(24y)(\Delta y)$. The distance from the origin up to the element is y. The distance from the element to the top of the tank is $8 - y$ feet. The work, Δw, required to lift all the water in the element to the top of the tank is

$$\Delta w = (8 - y)(\rho)(24y)(\Delta y) = (24\rho)(8y - y^2)\Delta y.$$

Adding up all the work for all the elements and taking the limit as the number of subintervals becomes infinite, we get the definite integral for the total work.

$$\text{work} = \int_0^8 (24\rho)(8y - y^2)dy = 2048\rho.$$

Substituting $\rho = 62.5$ lb/ft^3, we get

$$\text{work} = 128,000 \text{ ft} - \text{lbs.}$$

Note. No solution of a water pumping problem is complete unless the vertical axis is named on a drawing and a positive direction is clearly stated.

Example 3. Consider the region bounded by the graph of $y = (x - 3)^2$, $0 \le x \le 6$, and the line $y = 9$. This region is revolved about the y axis to form a container. The container is full of water. Find the work required to pump all the water to the level of the top of the container. All the numbers x and y are measured in feet.

Solution. A sketch of the graph of the region.

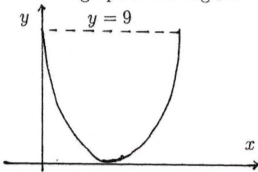

Divide the interval $0 \le y \le 9$ into small subintervals of length Δy. When the thin rectangle of height Δy is revolved about the y axis it forms a short cylinder (disk) with a hole in the middle. The volume is the volume of the disk minus the volume of the hole.

$$\Delta V = \pi(x_2)^2 \Delta y - \pi(x_1)^2 \Delta y.$$

The graph of $y = (x - 3)^2$ is a parabola. We need a formula for the $x = x_2$ on the right half of the parabola and a formula $x = x_1$ on the left half of the parabola.

$$(x - 3)^2 = y$$
$$x - 3 = \pm\sqrt{y}$$
$$x = 3 \pm \sqrt{y}$$

Values of x on the right half of the parabola are given by $x = 3 + \sqrt{y}$ and values of x on the left half of the parabola are given by $x = 3 - \sqrt{y}$. The

volume of the disk minus the hole is

$$\Delta V = \pi[3 + \sqrt{y}]^2 - \pi[3 - \sqrt{y}]^2$$
$$= 12\pi\sqrt{y}$$

The distance from this disk to the surface is $9 - y$. The total work is given by the definite integral

$$\text{work} = \rho \int_0^9 12\pi\sqrt{y}(9 - y)dy$$
$$= \frac{3888}{5}\pi\rho \text{ ft} - \text{lbs}.$$

EXAMPLE 4. The cross section of a large tank is a trapezoid, the bottom of the trapezoid is 10 ft long and the top is 26 ft long. The tank is 12 feet deep and is 15 ft long. The tank is full of water. The tank is emptied through a hose attached to the bottom of the tank. The end of the hose is 3 ft below the bottom of the tank. How much work is done in emptying the tank through this hose?

Solution. Choose up as positive. Let the horizontal axis, x, be along the bottom of the tank. Let the vertical axis, y, go up through the middle of the tank. Note that the origin is at the bottom of the tank in the middle. Divide the y axis into subintervals of length Δy. Consider a slice of water of height Δy and parallel to the ground. The volume of this slice is $\Delta v = 15(2x)\Delta y$. Using ρ as the density of water the weight of this slice is

$$15\rho(2x)\Delta y.$$

What is the direction of the force on this slice of water? What force? There is only on force being discussed and that is the force of gravity. We are finding the work done by the force of gravity. The magnitude of the force is the weight and the direction of this force is down. The force on this slice of water is

$$\Delta F = -[15\rho(2x)\Delta y].$$

The slanting side of the tank is part of a line. Let us find the equation of this line. In general any line through the point (x_1, y_1) with slope m is

$$y - y_1 = m(x - x_1).$$

Two points in this line are $(5,0)$ and $(13,12)$.

$$y - 12 = m(x - 13)$$
$$0 - 12 = m(5 - 13)$$
$$m = 3/2$$

The equation of the line is

$$y - 12 = 3/2(x - 13)$$
$$x = 2/3y + 5.$$

The force on this slice of water is

$$\Delta F = (-1)[\rho(15)(4/3y + 10)]\Delta y.$$

The distance that this slice of water moves when it leaks out of the tank is $y + 3$. The direction is down or negative. The displacement is $[-(y + 3)]$.

$$\text{work} = \text{force times displacement}$$
$$\Delta w = (-1)[\rho(15)(4/3y + 10)][-(y + 3)]\Delta y$$
$$= [\rho(15)(4/3y + 10)][y + 3]\Delta y.$$

Letting the number of subdivisions become infinite we get

$$\text{total work} = \int_0^{12} 15\rho[4/3y + 10][y + 3]dy$$
$$= \rho \int_0^{12} [20y^2 + 210y + 450]dy$$
$$= \rho[\frac{20}{3}y^3 + 105y^2 + 450y \,|_0^{12}]$$
$$= 32,040\rho$$
$$= 2,002,500 \text{ ft} - \text{lbs.}$$

If the force and displacement are in the same direction then the work is positive. This is work done by force of gravity as the water leaks from the tank.

Although it is not good we can solve this problem taking down as positive. Let us take the x axis along the top of the tank and y axis down the middle with y positive down. In this case

$$\Delta V = 15(2x)sy.$$

Two points on the line (part of slanting side) are $(13, 0)$ and $(5, 12)$. The equation is

$$x = 13 - (2/3)y.$$

The distance the slice of water moves is

$$15 - y.$$

The work done on one slice of water is

$$\Delta w = 15\rho[26 - (4/3)y](15 - y)dy$$

Letting the number of subintervals become infinite we get

$$\text{Total Work} = \rho \int_0^{12} (5850 - 690y + 20y^2)dy$$

$$= \rho[5850y - 345y^2 + \frac{20}{3}y^3 \, |_0^{12}]$$

$$= 32040\rho.$$

Exercises

1. A large tank is full of water. The tank is 5 feet wide, 8 feet long, and 4 feet deep. Find the work required to pump all of the water in the tank out an outlet 2 feet above the top of the tank. Use 62.5 lb/ft^3 as the weight density of water.

2. A large cylindrical tank is full of water. The tank is a circular cylinder with flat side down. The radius of the tank is 5 feet and the tank is 8 feet tall. Find all the work required to pump all the water out an outlet 4 feet above the top of the tank. Use 62.5 lbs/ft^3 as the weight density of water.

3. The cross section of a large tank is an isosceles triangle with one vertex on the ground. The top edge of the triangle that is parallel to the ground is 8 feet long. The two sloping sides are each $4\sqrt{5}$ feet long. The tank itself is 12 feet long and full of water. Find the work required to pump all the water to the level of the top of the tank.

4. The cross section of a large tank is a trapezoid. The short side of the trapezoid is 12 feet long and is the top of the tank. The long side of the trapezoid is 18 feet and is the bottom of the tank. The two sloping sides of the trapezoid are both 5 feet long. The tank itself is 15 feet long and is full of water. Find the work done in pumping all the water to a level 5 feet above the top of the tank.

5. A water tank is in the shape of a cone with the vertex down. (Just like a cone of ice cream). The cone is 15 feet deep and the radius of the top is 10 feet. The cone is full of water. Find the work done in pumping all the water to a level 4 feet above the top of the cone. Work = $3875\pi\rho$ ft-lbs.

6. Consider the region bounded by the graph of $y = x^2$, $0 \le x \le 5$, and the line $y = 25$. Note that the region is bounded on the left by the y axis. This region is revolved about the y axis to form a container. The container is full of water. Find the work required to pump all the water to the level of the top of the container. All the numbers x and y are measured in feet.

7. Consider the region bounded by the graph of $y = 2x^2$, $0 \le x \le 3$, the line $y = 18$, and the y axis which is in the first quadrant. The boundary curve is revolved about the y axis to form a container. The container is full of water. Find the work required to pump all the water to the level 4 feet above the top of the container. All the numbers x and y are measured in feet.

8. A large container of water is 40 ft long. Each cross section is an isosceles trapezoid with top 54 ft across, bottom 24 ft across, and height 20 ft. The tank is on the shore of a lake and is full of water. A hose is attached to a hole in the bottom of the tank and runs down hill 5 more feet to the surface of the lake. The tank is drained until empty through this hose. How much work is done?

351

6417 Mean Value Theorem for Integrals

Mean Value Theorem for Integrals. If $f(x)$ is continuous for $a \leq x \leq b$, then there is at least one number c with $a \leq c \leq b$ such that

$$\int_a^b f(x)dx = f(c)(b-a).$$

Proof. Let $m =$ the minimum value of $f(x)$ for $a \leq x \leq b$. Let $M =$ the maximum value of $f(x)$ for $a \leq x \leq b$, then

$$m \leq f(x) \leq M \text{ for } a \leq x \leq b.$$

Since $f(x)$ is continuous for $a \leq x \leq b$, there exists a value of x, say x_m, such that $f(x_m) = m$ and $a \leq x_m \leq b$. Since $f(x)$ is continuous for $a \leq x \leq b$, then there exists a value of x, say x_M, such that $f(x_M) = M$ and $a \leq x_M \leq b$. Next

$$m \leq f(x) \leq M \text{ for } a \leq x \leq b$$

$$\int_a^b m \, dx \leq \int_a^b f(x)dx \leq \int_a^b M \, dx$$

$$m(b-a) \leq \int_a^b f(x)dx \leq M(b-a)$$

$$m \leq \frac{1}{b-a} \int_a^b f(x)dx \leq M.$$

Since $f(x)$ is continuous for $a \leq x \leq b$, $f(x_m) = m$, and $f(x_M) = M$, there is some value of x such that $f(x)$ is equal to any number between m and M. In particular, there is a value of x, say $x = c$, such that

$$f(c) = \frac{1}{b-a} \int_a^b f(x)dx$$

or

$$\int_a^b f(x)dx = f(c)(b-a).$$

Example 1. Consider the function $3x^2 + 8x + 4$ and the interval $2 \le x \le 8$, find the number c that satisfies the Mean Value Theorem for Integrals.

Solution. We need the value of the integral

$$\int_2^8 (3x^2 + 8x + 4)dx = x^3 + 4x^2 + 4x \Big|_2^8$$

$$= 8^3 - 2^3 + 4(64 - 4) + 4(8 - 2) = 504 + 240 + 24 = 768.$$

The value of c is a solution of the equation

$$(3c^2 + 8c + 4)6 = 768$$

$$3c^2 + 8c - 124 = 0$$

$$c = \frac{-8 \pm \sqrt{64 - 4(3)(-124)}}{6} = \frac{-8 \pm \sqrt{1552}}{6}$$

$$c = 5.2326, \quad c = -7.8992.$$

The number $c = -7.8992$ does not satisfy the Mean Value Theorem because it is not between 2 and 8. The number c which satisfies the Mean Value Theorem is

$$c = 5.2326$$

Definition. The number

$$\frac{1}{b - a} \int_a^b f(x)dx$$

is the average value of $f(x)$ on the interval $a \le x \le b$.

Example 2. Find the average value of $f(x) = x^3 - 9x + 5$ on the interval $2 \le x \le 6$.

Solution. First

$$\int_2^6 [x^3 - 9x + 5]dx = \frac{x^3}{4} - \frac{9}{2}x^2 + 5x \Big|_2^6$$

353

$$= \frac{1}{4}[216 - 8] - \frac{9}{2}[36 - 4] + 5(6 - 2) = -72$$

The average value of $f(x) = x^3 - 9x + 5$ on the interval $2 \le x \le 6$ is

$$\frac{1}{6-2} \int_2^6 (x^2 - 9x + 5)dx = -\frac{72}{4} = -18.$$

We can also use the definite integral to find a unique antiderivative. Given the function $f(x)$ we can calculate

$$F(x) = \int_a^x f(t)dt,$$

where x is a variable and a is a constant. The expression $\int_a^x f(t)dt$ is usually called the indefinite integral of $f(x)$.

Basic Theorem. If $f(x)$ is continuous for all real numbers x, then the function

$$F(x) = \int_a^x f(t)dt$$

is an antiderivative of $f(x)$ and $F'(x) = f(x)$. We can express this as

$$\frac{d}{dx}\left[\int_a^x f(t)dt\right] = f(x).$$

The proof of this theorem uses the Mean Value Theorem for Integrals. We will not give the proof here.

Example 3. Let $f(t) = 6t^2 + 10t + 15$. Find $F(x)$ such that $F'(x) = f(x)$.

Solution.

$$\int_1^x f(t)dt = \int_1^x [6t^2 + 10t + 15]dt =$$

$$2t^3 + 5t^2 + 15t \Big|_1^x = 2x^3 + 5x^2 + 15x - 22 = F(x).$$

Note that $F'(x) = f(x)$. Also note that $2x^3 + 5x^2 + 15x - 22$ is only one of many antiderivatives of $6x^2 + 10x + 15$. We find other antiderivatives by replacing the lower limit one with other numbers. We usually express all antiderivatives as $2x^3 + 5x^2 + 15x + C$.

The indefinite integral is sometimes helpful in applications. Suppose an object is traveling along a straight line with velocity given by $v(t) = 6t^2 + 8t + 5$. How far does the object travel during the time integral $2 \leq t \leq u$?

$$\int_2^u (6t^2 + 8t + 5)dt = 2u^3 + 4u^2 + 5u - 42.$$

In order to express how far the object traveled during the time interval $2 \leq t \leq u$ we need this unique antiderivative.

The most useful feature of the function $F(x) = \int_a^x f(t)dt$ is that it provides an antiderivative to a function which is defined in sections.

Example 4. Let $f(x) = |x - 4|$, find $F(x)$ such that $F'(x) = f(x) = |x - 4|$ for all x.

Solution. Define $F(x)$ by

$$F(x) = \int_0^x |t - 4|dt.$$

If $x < 4$, then $t < 4$ and $t - 4 < 0$. This means that $|t - 4| = -(t - 4)$,

$$F(x) = \int_0^x (-t + 4)dt = -\frac{t^2}{2} + 4t \, \Big|_0^x = -\frac{x^2}{2} + 4x.$$

If $x \geq 4$, then we must consider $t \geq 4$. When $t \geq 4$, $t - 4 \geq 0$ and $|t - 4| = t - 4$. For $x \geq 4$,

$$F(x) = \int_0^4 (-t + 4)dt + \int_4^x (t - 4)dt =$$

$$-\frac{t^2}{2} + 4t \, \Big|_0^4 + \frac{t^2}{2} - 4t \, \Big|_4^x =$$

$$-8 + 16 + \frac{x^2}{2} - 4x - (8 - 16) = \frac{x^2}{2} - 4x + 16.$$

$$F(x) = \begin{cases} -\frac{x^2}{2} + 4x & x < 4 \\ \frac{x^2}{2} - 4x + 16 & x \geq 4. \end{cases}$$

Applying the Basic Theorem the function $F(x)$ is continuous and $F'(x) = f(x)$. A graph of $F(x)$ also shows that $F(x)$ is continuous. This is just one of the many antiderivatives of $f(x)$. Note that this is an antiderivative and may be used to find a definite integral.

$$\int_2^8 |x - 4| dx = F(x) \Big|_2^8 = \frac{8^2}{2} - 4(8) + 16 - \left[-\frac{4}{2} + 4(2) \right] = 10.$$

Example 5. Given the function

$$f(x) = \begin{cases} 4x - 3 & x < 2 \\ 3x^2 - 2x - 3 & x \geq 2 \end{cases}$$

find $F(x)$ such that $F'(x) = f(x)$.

Solution. Note that $f(x)$ is continuous for all x. In particular, $f(2) = 5$. We find the function $F(x)$ in two parts. First, for all $x < 2$

$$F(x) = \int_0^x (4t - 3) dt = 2t^2 - 3t \Big|_0^x = 2x^2 - 3x.$$

If $x \geq 2$, then

$$F(x) = \int_0^x f(t) dt = \int_0^2 (4t - 3) dt + \int_2^x (3t^2 - 2t - 3) dt =$$

$$2t^2 - 3t \Big|_0^2 + t^3 - t^2 - 3t \Big|_2^x =$$

$$(8 - 6) + x^3 - x^2 - 3x - (8 - 4 - 6) = x^3 - x^2 - 3x + 4.$$

$$F(x) = \begin{cases} 2x^2 - 3x & x < 2 \\ x^3 - x^2 - 3x + 4 & x \geq 2. \end{cases}$$

This function $F(x)$ is continuous and $F'(x) = f(x)$. Just to be clear $F'(2)$ exists and $F'(2) = f(2) = 5$.

Exercises

1. Consider the function $f(x) = 6x^2 - 4x + 5$ and the interval $1 \leq x \leq 6$. Find the number c that satisfies the Mean Value Theorem for Integrals.

2. Consider the function $f(x) = \sin x$ and the interval $0 \le x \le \pi$. Find the number c that satisfies the Mean Value Theorem for Integrals for this function on this interval.

3. Find the average value of $f(x) = 8x^3 + 6x^2 + 5$ on the interval $2 \le x \le 6$.

4. Find the average value of $f(x) = 6x - x^2$ on the interval $2 \le x \le 8$.

5. Let $f(x) = |x - 3|$. Find $F(x)$ such that $F'(x) = |x - 3|$ for all x.

6. Let $f(x) = \begin{cases} -2x + 10 & x < 3 \\ x^2 - x - 2 & x \ge 3. \end{cases}$ Find $F(x)$ such that $F'(x) = f(x)$ for all x. Note that $f(x)$ is continuous for all x.

7. Let $f(x) = \begin{cases} -x^2 - 7x + 21 & x < 2 \\ x^2 - 2x + 3 & x \ge 2. \end{cases}$ Find $F(x)$ such that $F'(x) = f(x)$ for all x. Note that $f(x)$ is continuous for all values of x and that $f(2) = 3$.